基于金属有机骨架材料构建高效光催化剂的研究

刘 琪/著

中国矿业大学出版社
·徐州·

内 容 提 要

本书为了突破单一半导体光催化剂性能上的限制,基于催化剂的载流子分离效率、CO_2 吸附活化、产物选择性等问题,采用金属有机骨架(MOFs)作为反应前驱体或复合单元,设计制备了一系列界面可控的复合结构体系,系统研究了其光催化性能,揭示了复合体系光催化剂的反应机理。

本书可供相关专业的研究人员借鉴、参考,也可供广大教师教学和学生学习使用。

图书在版编目(CIP)数据

基于金属有机骨架材料构建高效光催化剂的研究/
刘琪著. ─徐州:中国矿业大学出版社,2022.8
ISBN 978-7-5646-5493-1

Ⅰ. ①基… Ⅱ. ①刘… Ⅲ. ①金属材料─有机材料─
应用─光催化剂─研究 Ⅳ. ①O643.36

中国版本图书馆 CIP 数据核字(2022)第 151754 号

书　　名	基于金属有机骨架材料构建高效光催化剂的研究
著　　者	刘　琪
责任编辑	何晓明　何　戈
出版发行	中国矿业大学出版社有限责任公司
	(江苏省徐州市解放南路　邮编 221008)
营销热线	(0516)83885370　83884103
出版服务	(0516)83995789　83884920
网　　址	http://www.cumtp.com　E-mail:cumtpvip@cumtp
印　　刷	苏州市古得堡数码印刷有限公司
开　　本	787 mm×1092 mm　1/16　印张 9.25　字数 181 千字
版次印次	2022 年 8 月第 1 版　2022 年 8 月第 1 次印刷
定　　价	38.00 元

(图书出现印装质量问题,本社负责调换)

前　言

随着人类日益增长的能源需求与能源日益短缺矛盾的加剧，新能源尤其是太阳能的开发利用显得尤为重要。光催化反应条件温和，能将太阳能直接转变为化学能，在太阳光的作用下，光催化剂产生的光生电子和空穴可以将水分解产氢、将 CO_2 还原成碳氢燃料、降解有机污染物、还原重金属离子等，具有节能、绿色、可持续和低成本等特点，为碳达峰和生态文明建设提供强有力的支撑。然而，光催化仍然面临着太阳能转换效率低的问题。高效、低成本地开发先进的光催化剂是一个巨大的挑战。

金属有机骨架材料（Metal-Organic Frameworks，MOFs）由于结构上的多样性、多孔性、可剪裁性以及具有超高比表面积、优异的气体吸附性能等特性，在光催化领域引起了广泛的关注。本书以MOFs材料为主题材料，首先综述了该类材料在各类光催化反应中的应用，围绕光的有效利用、光生载流子分离和反应物吸附等方面，以MOFs材料为基础介绍了笔者近几年在这个领域的几项工作。第2章和第3章主要从反应物的高吸附入手，设计了两种MOFs与半导体材料组成的复合光催化剂；第3~6章以MOFs为模板或反应物，设计了4种独特结构的催化剂材料。

本书由安徽工程大学刘琪独立撰写，在撰写过程中研究生许淼、周贝贝、李春晓、武恒、唐亚文、郑梦娇、刘子义和来龙杰做了大量的工作。安徽工程大学孙宇峰教授对本书的撰写也十分关心和支持,在此一并感谢。

　　本书的撰写是笔者新的尝试,由于水平有限、经验不足,书中疏漏之处在所难免,恳切希望广大读者提出宝贵意见。

著　者
2022 年 6 月

目 录

第1章 绪论 ·· 1
   1.1 引言 ·· 1
   1.2 光催化及光催化材料 ··· 2
   1.3 金属有机骨架材料 ·· 8
   1.4 半导体/金属有机骨架复合光催化剂 ·························· 11
   1.5 MOFs 衍生光催化材料 ·· 20
   参考文献 ·· 27

第2章 TiO_2/ZIF-8 复合材料的制备及其光催化性能研究 ············· 43
   2.1 引言 ··· 43
   2.2 实验部分 ·· 44
   2.3 结果与讨论 ·· 46
   2.4 本章小结 ·· 55
   参考文献 ·· 55

第3章 ZIF-8/Zn_2GeO_4 纳米棒的制备及其光催化还原 CO_2 研究 ········ 61
   3.1 引言 ··· 61
   3.2 实验部分 ·· 62
   3.3 结果与讨论 ·· 64
   3.4 本章小结 ·· 75
   参考文献 ·· 75

第 4 章　多孔 ZnO 纳米片薄膜的制备及其在染料敏化太阳能电池中的应用研究 ························· 80
 4.1　引言 ························· 80
 4.2　实验部分 ························· 81
 4.3　结果与讨论 ························· 82
 4.4　本章小结 ························· 93
 参考文献 ························· 93

第 5 章　Fe^{3+} 掺杂 TiO_2 八面体的制备及其光催化还原 CO_2 研究 ························· 98
 5.1　引言 ························· 98
 5.2　实验部分 ························· 99
 5.3　结果与讨论 ························· 100
 5.4　本章小结 ························· 109
 参考文献 ························· 109

第 6 章　Cr_2O_3/C@TiO_2 复合材料的制备及其光催化产氢性能研究 ························· 115
 6.1　引言 ························· 115
 6.2　实验部分 ························· 117
 6.3　结果与讨论 ························· 117
 6.4　本章小结 ························· 136
 参考文献 ························· 136

第1章 绪　　论

1.1 引言

能源危机和环境污染问题是当今世界亟待解决的课题。不可再生能源因过度开采和无节制的使用而面临枯竭;不断增长的石化能源消耗,使得大气中 CO_2 浓度逐年上升,引起了一系列严重的环境问题。同时,人类日常生活与生产的过程中,废水、废渣等污染物未经处理直接排放到环境中,造成严重的环境污染,严重危害人类的日常生活及健康。因此,寻找替代能源和修复生态环境成为各国科学家们和政府的重大研究课题之一。1972 年,Fujishima 等[1]首次使用 TiO_2 电极进行光电催化分解水。随后,Carey 等[2]成功地将 TiO_2 用于光降解水中多氯联苯;Halmann[3]以 P 型半导体 GaP 作为光电极,将 CO_2 还原得到甲醛、甲酸和甲醇。这一系列重要研究成果使基于半导体的光催化技术展现出良好的应用前景。

迄今为止,已报道的光催化剂大多数为半导体材料,如 ZnO[4-6]、WO_3[7-9]、Fe_2O_3[10-11]、TiO_2[11-15]、C_3N_4[16-19]、$ZnGa_2O_4$[20-21]、Zn_2GeO_4[22-26]等。半导体材料光催化反应主要包含三个基本过程:① 半导体材料吸收能量大于禁带宽度的光子受激发产生光生电子-空穴对;② 光生电子-空穴对有效分离,迁移到半导体表面的活性位点;③ 在半导体表面发生氧化-还原反应。

所以要实现高效光催化反应,应围绕着太阳光捕获、反应物吸附和活化、电荷分离和传输来设计构建高活性的光催化剂。太阳光捕获方面,利用能带理论调节带隙和带边位置来有效利用太阳能,采用离子掺杂、固溶体、异质结和光敏化等方式开发了一批可见光响应光催化半导体材料。然而,目前传统的半导体材料的太阳能转换效率仍然很低,最高只有 1‰量级。根本原因是:① 低的比表面积,表面活性位点少;② 电子-空穴对复合严重,寿命短。因此,提升半导体材料的比表面积、增加活性位点、抑制电子-空穴对的复合、促进电子-空穴对的有效分离成为提升半导体材料催化活性的重要途径[27]。

1.2 光催化及光催化材料

1.2.1 光催化

传统的光催化剂多为半导体材料,具有与金属和绝缘体不同的能带结构,即由低能价带、高能导带和带隙构成半导体的非连续性能带结构。价带和导带分别指一系列满带最上面的满带与一系列空带最下面的空带。价带与导带之间存在一个禁带,即为带隙。禁带宽度即为它的带隙能(E_g),其决定半导体的响应波长($E_g = h\nu = hc/\lambda$)。半导体光催化剂的光催化机理如图 1-1 所示。

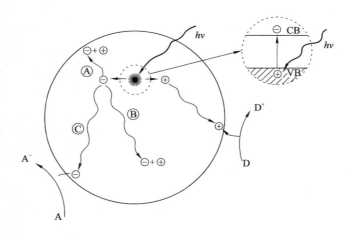

图 1-1 光催化基本原理图[28]

下面以 TiO_2 为例来说明整个光催化的基本过程。TiO_2 的带隙能(E_g)一般为 3~3.2 eV。当光子的能量 $h\nu \geqslant E_g$ 时,电子受激发由价带跃迁至导带上,因此,在价带上生成具有强氧化能力的空穴(h^+),在导带上生成具有还原能力的电子(e^-),在空间电场的作用下,电子和空穴通过扩散的方式分别向 TiO_2 的表面迁移。在迁移的过程中,光生电子和空穴会在内部发生复合(图 1-1 中的 B 过程)和表面发生复合(图 1-1 中的 A 过程)。成功迁移至表面且没在表面发生复合的空穴(图 1-1 中的 D 过程)和电子(图 1-1 中的 C 过程)是光催化反应的活性物质。活性物质与吸附在 TiO_2 表面的反应物发生氧化反应或还原反应,同时,空穴还可与表面吸附的水或 OH^- 反应生成氧化性更强的·OH 自由基,电子又可与表面吸附的 O_2 发生作用,生成具有极强还原能力的·O_2^-。这些高活性基团不仅能在一定程度上使多种有机物彻底矿化分解成为水和二氧化碳等无机物小分子,还能降解部分有毒有害的无机物,使其生成无毒无害的无机物。

1.2.2 光催化材料

(1) 新型光催化材料不断涌现

金属氧化物是最为广泛研究的光催化还原 CO_2 材料,TiO_2、ZnO、Cu_2O、ZrO_2、WO_3 等是广泛研究的光催化材料。近年来,多元复杂氧化物也表现出了良好的光催化活性,如 $SrTiO_3$、$K_2Ti_6O_{13}$、$SrLa_4Ti_4O_{15}$、$InNbO_4$、$NaNbO_3$、$InTaO_4$、$BiVO_4$、Zn_2GeO_4 等。南京大学周勇教授课题组相继报道了 $BiVO_4$[29-30]、Bi_2WO_6[31-32]、Zn_2GeO_4[24,26]、$In_2Ge_2O_7$(En)[33]、Zn_2SnO_4[34-35]、$Fe_2V_4O_{13}$[36] 和 $Na_2V_6O_{16}$[37] 等多元氧化物光催化材料。此外,金属硫化物(CdS、Bi_2S_3、ZnS、MnS、Cu_2ZnSnS_4、$ZnIn_2S_4$)、磷化物(InP、GaP)以及 SiC、Si 也表现出良好的光催化性能。

(2) 能带工程在光催化设计中广泛应用

光催化剂不仅要具有合适的带隙大小,还应具有适当的能级位置(图 1-2),以光催化还原 CO_2 为例,导带的电位应比 CO_2/碳氢化合物的还原电位更负;同时,价带的电位应比 O_2/H_2O 氧化电位更正。通过能带剪裁(掺杂、敏化、固溶等),调节光催化剂的带隙大小、导价带能级位置,从而调控光催化性能。Khan 等[38]制备的 N 掺杂 TiO_2,其可见光吸收范围可扩展到 500 nm,将 CO_2 的还原为 CH_4,提高了光催化性能。N 掺杂 Ta_2O_5 或 $InTaO_4$ 也同样增加了可见光的吸收,提高了光催化 CO_2 的还原活性[39-42]。同样,Cr 掺杂使得 TiO_2 光电化学分解水性能获得了提高[43]。用 CdS、Bi_2S_3、CdSe、PbS、AgBr、CuI 等敏化 TiO_2 都能提高光催化性能。南京大学邹志刚教授用 CuI 染料分子敏化 TiO_2,在可见光下将 CO_2 有效地还原为 CH_4[44]。固溶体往往表现出比单一组分更好的光催化还原 CO_2 活性。南京大学周勇教授研究发现,$Zn_{1.7}GeN_{1.8}O$ 固溶体中 N 2p 和 Zn 3d 的互斥使价带顶提高,带隙减小,在可见光下还原 CO_2 为 CH_4 的表观量子产率提高到 0.024%[26]。窄带隙固溶体 $ZnIn_2S_4$(<2 eV)具有较高的可见光催化还原 CO_2 活性,与 $BiVO_4$ 复合后极

大地提高了 CH_4 的产率[45]。

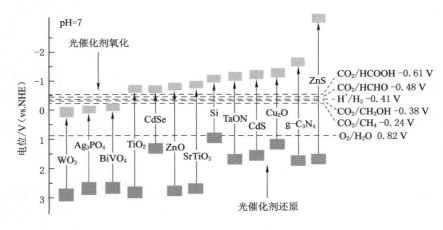

图 1-2　常见半导体材料的导带和价带位置

(3) 重视光催化材料的构-效关系

相比多晶材料,纳米光催化剂(纳米晶、纳米线、纳米带等)往往由单晶组成,晶界或缺陷较少,减少了电子-空穴复合中心。粒径为 14 nm 的锐钛矿 TiO_2 纳米晶具有良好的光吸收、较大的比表面积和有效的光生电荷分离,具有良好的光催化活性,可将 CO_2 还原为 CH_4 和 CH_3OH[46]。一维纳米材料为光生电荷提供了传输通道,有利于光生电子与空穴分离。因此,一维 Zn_2GeO_4 纳米带[24]、$In_2Ge_2O_7(En)$[33] 纳米线和 $Cd_2Ge_2O_6$[23] 纳米线都具有较高的光催化还原 CO_2 性能。

选择性暴露晶面可以调控能带结构、表面能、表面活性位点、活性物种吸附等,成为调控光催化性能的另一个有效途径。CH_4 只在(100)暴露面 TiO_2 光催化还原 CO_2 产生,此外也含有 CH_3OH 产物。然而 CH_4 和 CH_3OH 在(110)暴露面 TiO_2 的产率小于(100)暴露面 TiO_2 的产率,主要是因为(100)暴露面 TiO_2 具有更高的 Ti/O 比,能提供更多的反应位点[47]。周勇教授与其合作者合成了{100}晶面 $ZnGa_2O_4$,有效地将 CO_2 还原为 CH_4[20-21]。中科院沈阳金属所刘岗、成会明等合成了{010}晶面锐钛矿 TiO_2、{002}晶面 WO_3,将 CO_2 还原为 CH_4[48-49]。

此外,具有纳米空心结构、多孔结构、分级结构、金属有机骨架结构的光催化材料均具有较大的比表面积,显示出比普通颗粒更好的光催化性能。研究发现,与块材相比,介孔 $ZnGa_2O_4$ 光催化还原 CO_2 具有更高的光活性[21]。日本国家材料研究所叶金花教授报道了 $ZnGe_2O_4$ 介孔光催化材料[50],随后又制备了 N 掺杂的 $ZnGa_2O_{4-x}N_x$ 介孔光催化材料,光催化还原 CO_2 活性从紫外光扩展到了可见光范围[51]。福州大学付贤智、李朝辉等报道了金属有机骨架配合物 NH_2-MIL-125(Ti)和 NH_2-UiO-66(Zr)在可见光下将 CO_2 还原为 $HCOO$[52-53]。

(4) 助催化剂被广泛应用

除离子掺杂外,另一种提高电子-空穴对寿命和将光吸收边扩展到可见区域的有效方法是在催化剂表面沉积贵金属。在 Pt、Rh、Pd、Cu、Ag、Au 等金属粒子助催化剂与半导体光催化剂之间形成肖特基势垒,金属粒子的导带底在半导体光催化剂导带底之下,促进了光生电子空穴的分离与传输,有助于光催化还原 CO_2 性能的提高。TiO_2 担载 Pt(Rh、Pd、Cu、Ru、Au 等)后可以更加有效地将 CO_2 还原为碳氢燃料。与金属助催化剂不同,RuO_2 捕获空穴参与氧化反应,与 Pd 共担载到 TiO_2 后,还原 CO_2 为 $HCOO^-$,提高了光催化性能[54]。$InTaO_4$ 担载 NiO 后,可见光下选择性还原 CO_2 为 CH_3OH[55]。助催化剂亦能有效阻止逆反应,如 $BaLa_4Ti_4O_{15}$ 担载 Ag 后,提高了光催化还原 CO_2 为 CO 的效率[56]。

此外,Ag、Au 等贵金属纳米颗粒产生的表面等离子体共振效应拓宽光响应范围和提高光催化还原 CO_2 性能引起了人们的关注。TiO_2 负载 Au 纳米粒子后,在可见光区(532 nm)有明显的等离子体共振峰,具有很宽的光谱吸收范围,光催化还原 CO_2 的活性提高了近 24 倍[57]。

石墨烯、石墨碳、碳纳米管也可作为光生电荷的传输介质,是应用于光催化还原 CO_2 的另一类助催化剂。周勇教授研究组发现石墨烯-TiO_2 光催化剂中 d-π 电子轨道交叠(Ti-O-C 能带)有助于光生电子从 TiO_2 传输到石墨烯,具有较高的催化活性,可以将 CO_2 还原为 CH_4 和 C_2H_6[6]。

(5) 复合光催化材料逐渐兴起

纳米复合催化材料不仅可以扩大光响应范围,还可以增大比表面积、增加活性位置及促进光生电荷的分离与迁移,提高光催化性能。Bi_2S_3/CdS、CeO_2/TiO_2、$CdSe/TiO_2$、Bi_2S_3/TiO_2 等异质结光催化材料的光生电荷在内建势场的作用下从一种光催化剂的导带(价带)跃迁到另一种光催化剂的导带(价带),促进了光生电荷的分离与传输,有利于光催化还原 CO_2 效率的提高。CuO/Fe_2O_3、Cu_2O/TiO_2、CuO/TiO_2 等 PN 结光催化材料在内建电场的作用下光生电荷有效分离,提高了光催化还原 CO_2 的性能。由于同质异构材料能带结构的差异,同质异构结也能有效减小光生电荷的复合概率。锐钛矿 TiO_2

的导带能级比金红石 TiO_2 的更负,锐钛矿 TiO_2 的光生电子易迁移至金红石 TiO_2 的导带,促进了光生电荷的分离。因此,金红石 TiO_2 纳米颗粒与锐钛矿 TiO_2 纳米棒组成的同质结提高了光催化还原 CO_2 的性能[58]。

在半导体表面构建异质结是另外一种能有效强化光催化效率的方法,表面异质结能够有效促进载流子的界面分离。构建高效界面、研究其光生电荷迁移机制仍是光催化领域的重要课题。典型的异质结可分为两大类,具体如图1-3所示[59]:① 金属/类金属-半导体异质结(肖特基结)。② 半导体-半导体异质结,包括 I 型异质结、II 型异质结及 Z 型异质结等。II 异质结能够实现光激发条件下电子-空穴的有效界面转移和空间分离。Z 型异质结是自然界植物光合作用光反应阶段所采用的电荷转移方式。为了模拟自然光合作用,提高光催化剂的氧化还原能力,巴德在1979年提出了传统 Z 型异质结光催化剂结构,由于电子转移过程在图中构成英文字母 Z 的形状,因而称之为 Z 型。Z 型反应一个重要的特点就是氧化还原中间体(电子介体)的存在实现了电子和空穴的有效传递和分离,其独特之处最先应用于光催化分解 H_2O 体系的设计。

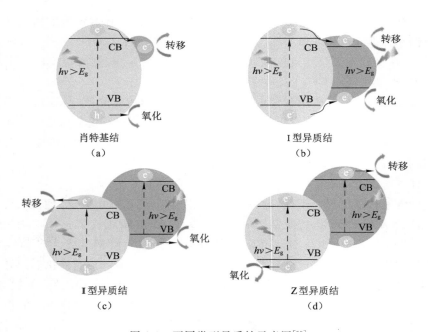

图1-3 不同类型异质结示意图[59]

Z型全分解水体系中的两种光催化剂以及电子介体往往悬浮在液体中,影响光的高效吸收,电子亦不能进行有效传递;在一定条件下,电子介体可能会被光生电子(或空穴)选择性地还原(或氧化),或与 $H_2(O_2)$ 发生反应;电子介体还可能会引起一些逆反应;有颜色的电子介体还可能会吸收一部分入射光。因此,在光催化水分解体系中,人们开发出无电子介体的全固态 Z 型反应体系 $CdS/Au/ZnO$、$CdS/Au/TiO_2$、$WO_3/W/Si$、TiO_2/C_3N_4、$SrTiO_3/RGO/BiVO_4$(RGO:还原氧化石墨烯)等[60-63],如图 1-4 所示。Au、W、RGO 成为光生电荷传输通道,使光生电子空穴有效分离与迁移。

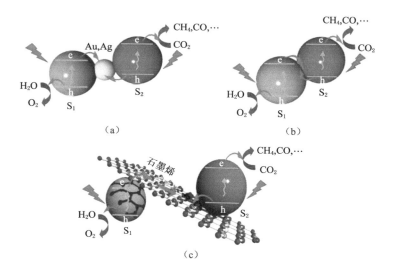

图 1-4　固态 Z 型光催化还原 CO_2 反应

南京大学周勇教授开发出可见光响应($\lambda \geqslant 420$ nm)Z 型光催化还原 CO_2 体系 $Fe_2V_4O_{13}/RGO/CdS$。实验研究表明,光催化氧化还原反应能够在常温常压下持续进行[64]。CdS 的光生空穴将 H_2O 氧化释放出 O_2,$Fe_2V_4O_{13}$ 的光生电子将 CO_2 还原为 CH_4。此外,$Fe_2V_4O_{13}$ 的光生空穴被 RGO 传来的 CdS 光生电子复合,促进了光生电荷的有效分离和传输,大幅提高了光催化性能(图 1-5)。除了传统的无机物半导体材料,金属有机骨架材料结构上的多样性、多孔性、可剪裁性以及超高比表面积、优异的气体吸附性能等特性,使其在光催化领域备受关注。

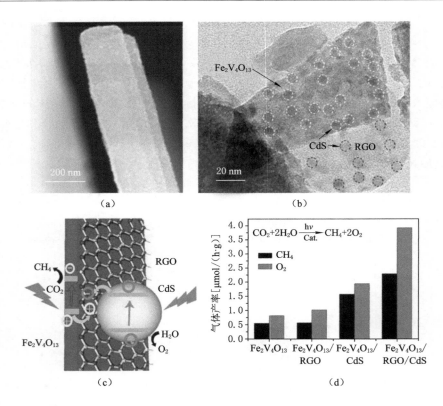

图 1-5 $Fe_2V_4O_{13}$/RGO/CdS 的 SEM 图、TEM 图、反应示意图以及
不同催化剂 CH_4 与 O_2 产率柱状图[64]

1.3 金属有机骨架材料

金属有机骨架材料(Metal-Organic Frameworks,MOFs)是一类由金属离子/簇和有机配体相连接组成的结晶性多孔材料,具有大小可调的孔道以及较高的比表面积。由于结构上的多样性、多孔性、可剪裁性以及超高比表面积、优异的气体吸附性能等特性,其广泛应用于化学传感器、生物医学、气体吸附和分离、质子传导及非均相催化等领域。

1.3.1 MOFs 材料特性

由于 MOFs 材料是由有机组分和无机组分相结合构成,因此,与传统的多孔材料相比,具有如下优点:① 种类多。可作为配体的有机物种类很多,包括咪唑酯、吡啶类、酚类、胺类、磷酸酯等,目前已经报道的 MOFs 结构有 20 000 多种,但仍是冰山一角,理论上可合成的数量是无限的。② 功能性强。通过选择多齿有机配体与金属离子的种类制备出不同的 MOFs 材料。同时,可通过引入不同的功能基团,进行性能调控,制备功能性 MOFs 材料。③ 孔隙率大和高的比表面积以及晶体密度小。合成材料的孔道吸附的溶剂分子,经活化处理后,溶剂分子被去除,形成具有多孔结构的骨架。④ 孔尺寸可调控性强。骨架孔隙的大小主要是由有机配体的长度和金属离子的大小所决定的。⑤ 生物相容性。通过采用生物分子作为有机配体与生物相容性的金属离子,从而合成具有生物相容性的生物 MOFs 材料(bio-MOFs)。这些特点使得 MOFs 材料在多个领域有潜在的应用价值,且已被应用于催化、气体存储和磁学等领域。其中,在光催化领域里,MOFs 材料通常应用于染料降解、小分子催化转化、裂解水制氢和 CO_2 还原等方向。

1.3.2 MOFs 材料在光催化中的应用

传统半导体光催化剂面临的主要问题包括可见光响应差、光生载流子复合严重、活性位点暴露受限等。MOF 具有类似半导体的行为,其具有结构明确及高比表面积等特性,引起了人们对其在光催化和研究领域的广泛兴趣。

(1) MOFs 材料光催化剂

很多 MOFs 具有半导体特性,在光照下可以产生光生电子和空穴,Usman 等[65]通过研究 MOFs 的能带结构,认为 MOFs 是未来的窄带隙材料,具有可见光响应。2007 年,有学者将 MOF-5 引入光催化领域进行苯酚降解,此后,以 IRMOFs、ZIF、MIL、UiO 系列等具有类半导体特性的 MOFs 在光催化领域迅速发展。2009 年,Kataoka 等[66]率先报道了多孔 MOF$[Ru_2(p\text{-}BDC)_2]_n$ 用于光催化分解水产 H_2。2012 年,福州大学李朝晖教授等第一次发现 Ti 基 MOF 材料$[NH_2\text{-}MIL\text{-}125(Ti)]$在可见光下可以光催化还原 CO_2 生成甲酸[67]。研究者们基于 MOFs 灵活和稳定的结构,通过合成后修饰的方法将活性位点引入 MOFs 中,或者通过化学反应构筑 MOFs 中的反应活性位点,进一步扩展了 MOFs 光催化剂的种类。中国科学技术大学江海龙教授等制备了一种锆基 MOF 材料(PCN-222),通过有效整合 CO_2 捕获与可见光光催化双功能于一体,实现了从 CO_2 到甲酸根离子的高效/高选择性转化[68]。Lee 等[69]利用含有(Zr/Ti)的 UiO-66 在可见光下将 CO_2 光催化还原成 HCOOH。NH_2-UiO-66(Zr)[70]、Fe-MOFs[71]、UiO-67[72]等 MOFs 也被证明具有光催

化还原 CO_2 活性。众所周知，CO_2 在光催化剂表面的活化是其光催化还原反应速度的控制步骤[73]。中国科学院国家纳米科学中心唐智勇等研究发现，MOFs 上的配位不饱和金属位点对 C=O 双键具有强的吸附和活化作用[74]。采用 Co-ZIF-9 为助催化剂，MOFs 优良的 CO_2 吸附性能和其有机体对 CO_2 的强稳定效应，促进了 CO_2 分子的活化，从而提高了催化剂光催化还原 CO_2 的活性[75]。

当前，通过调控钛 MOFs 价带能级或活性位点，构筑高效的光催化剂仍是光催化领域的研究热点。比如，通过合成后修饰方法制备的氮化碳@MIL-125，构筑了对 Cr(Ⅵ)还原和染料矿化具有高催化活性的新型光催化体系[76]。原位合成的 In_2S_3@MIL-125 核壳结构微颗粒可以高效去除废水中的抗生素[77]。虽然 MOFs 催化剂具有高的比表面积和丰富的活性位点，但是其结构尚不能完全满足光催化反应关于光吸收、电荷迁移、吸附等各方面的要求。

（2）金属纳米粒子与 MOFs 复合

贵金属纳米粒子通常可作助催化剂，吸收电子，加速电子和空穴的分离，从而提升光催化效率。Horiuchi 等[78]将 Pt 纳米粒子通过光沉积的方法负载在 Ti-MOF-NH_2 上，得到的 Pt/Ti-MOF-NH_2 复合光催化材料产氢性能优异。如图 1-6 所示，他们提出光催化反应中，激发态氨基对苯二羧酸配体产生的电荷可以有效转移到 Ti-O 簇的导带上，再转移到 Pt 上进行光解水制氢。Pt 作为共催化剂有效提高了光催化活性。

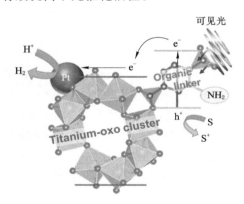

图 1-6　Pt/Ti-MOF-NH_2 光催化反应的示意图[78]

除此之外，贵金属 Pt 颗粒与 MOFs 颗粒的相对位置会影响 MOFs 光催化剂光生电子-空穴的分离和转移。中国科学技术大学江海龙等分别合成了负载于 MOFs 孔道内和表面的 Pt@UiO-66-NH₂ 和 Pt/UiO-66-NH₂(图 1-7)，其产氢效率均高于单一的 UiO-66-NH₂，且 Pt@UiO-66-NH₂ 产氢速率最高，这是由于该结构的电子传输距离的明显缩短以加速电子-空穴对的分离[79]。同时，江海龙等还制备了 Pt@MIL-125/Au 和 Pt/MIL-125/Au 两种复合材料，将 MOFs 与 Pt 形成的肖特基结和 Au 的等离子共振效应相结合，不仅可将光吸收拓展至可见光，还促进电子-空穴分离，并且 Pt 在 MOFs 孔道内的 Pt@MIL-125/Au 依然具有最高的光催化产氢速率[80]。

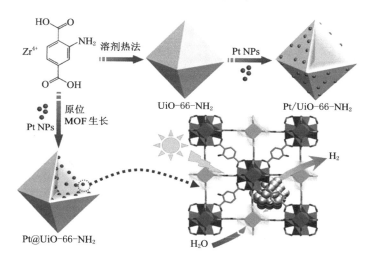

图 1-7　Pt@UiO-66-NH₂ 和 Pt/UiO-66-NH₂ 示意图[79]

1.4　半导体/金属有机骨架复合光催化剂

与传统的光催化剂相比，MOFs 材料虽然具有优异的吸附和活化性能，但其光催化效率非常低，主要原因是电荷分离效率低，电子-空穴对复合严重，这限制了其在光催化反应中的应用。将无机半导体材料与具有大比表面积且分子剪裁与修饰功能优良的 MOFs 材料结合，首先利用 MOFs 的多孔结构和大比表面积促进半导体纳米粒子的分散，获得更多的反应活性中心；其次利用 MOFs 实现反应物在光催化剂表面的有效富集和活化，而且可以促进电子-空穴对的有效分离，势必提升光催化剂活性。因此，合理设计半导体/MOFs 复合材料是制备新型高效光催化剂的有效途径。

1.4.1 TiO$_2$/MOFs 复合材料

TiO$_2$ 因其成本低、无毒、热稳定性好而被广泛使用于光催化领域[81-83]。然而 TiO$_2$ 由于具有较低的比表面积和孔隙率、较高的电子-空穴复合率,这限制了其光催化活性。提高 TiO$_2$ 的孔隙率和比表面积,降低光生载流子复合率是 TiO$_2$ 基光催化剂设计的关键。将 MOFs 与 TiO$_2$ 结合为制备新型光催化剂提供了新的策略[84]。近年来,很多研究团队通过 MOFs 与 TiO$_2$ 的复合,形成具有比纯 TiO$_2$ 更好的光催化活性的复合材料。中国科学技术大学熊宇杰教授等合成了 Cu$_3$(BTC)$_2$@TiO$_2$ 光催化剂,通过超快吸收光谱证实电子能从 TiO$_2$ 壳传递到 Cu$_3$(BTC)$_2$ 核上(图 1-8),使吸附到 MOF 材料上的 CO$_2$ 分子活化,从而使得该复合光催化剂在光催化还原 CO$_2$ 中表现出优越的选择性和活性[85]。Chang 等[86]合成了 TiO$_2$@MIL-53 复合材料,实验结果表明,TiO$_2$@MIL-53 核壳复合材料有较好的吸附和光催化降解作用。此外,高温活化 TiO$_2$@MIL-53 显示出较强的吸附能力。传统的半导体光催化剂对 CO$_2$ 吸附性能差,极大地制约了其光催化还原 CO$_2$ 活性。笔者以 ZIF-8 为吸附材料、Zn$_2$GeO$_4$ 为半导体光催化剂,在 Zn$_2$GeO$_4$ 纳米棒表面沉积 ZIF-8 颗粒,构建了新型 ZIF-8/Zn$_2$GeO$_4$ 复合光催化剂,借助 ZIFs 组元提高催化剂的 CO$_2$ 吸附量,实现了催化剂表面 CO$_2$ 的富集和高效还原。同样以 ZIF-8 为吸附材料、TiO$_2$ 为半导体光催化剂,在 TiO$_2$ 介孔小球表面沉积 ZIF-8 颗粒,构建了新型 ZIF-8@TiO$_2$ 核壳结构复合光催化剂,引入 ZIF-8 后,催化剂光催化还原 Cr(VI) 为 Cr(III) 的活性得到了大幅度提升[22,87]。

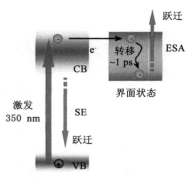

图 1-8 Cu$_3$(BTC)$_2$@TiO$_2$ 光生电荷传输示意图[85]

Zeng 等[88]通过超声化学法在 TiO_2 纳米纤维表面复合 ZIF-8,获得了 TiO_2/ZIF-8 复合材料(图 1-9)。光催化实验结果表明,ZIF-8 与 TiO_2 纳米纤维之间形成了 N—Ti—O 键,提升了复合材料的热稳定性、结晶度,促进了光生电子-空穴的有效分离,从而大幅度提高了其光催化降解 RhB 的能力。而且,与 ZIF-8 复合后,TiO_2/ZIF-8 复合材料的光吸收扩展至可见光。Wang 等[89]通过水热法获得 CPO-27-Mg/TiO_2 纳米复合材料。结果表明,由于 CPO-27-Mg 较强的 CO_2 吸附能力和存在的大量开放的碱金属活性位点,该复合材料的 CO_2 光催化还原能力较纯 TiO_2 有显著的提升。

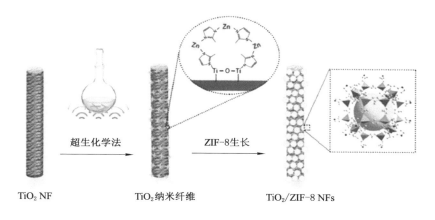

图 1-9 TiO_2/ZIF-8 复合材料制备示意图[88]

Sheng 等[90]通过多步自组装的方法在中空 TiO_2 纳米粒子表面生长了 MIL-101(Cr),形成 TiO_2@MIL-101(Cr)双层中空颗粒,并考察了复合材料的催化活性。结果表明,由于制备的复合材料的高比表面积和 MIL-101(Cr)对 H_2S 气体高的吸附与脱附能力,TiO_2@MIL-101(Cr)复合材料光催化 60 min 时,H_2S 转化率为 90.1%,与中空 TiO_2 纳米粒子及 TiO_2 相比,催化性能分别提升了 31% 和 114%。此外,TiO_2@MOF-5[91]、HKUST-1@TiO_2[92]、TiO_2@MIL-100[93]等复合光催化剂由于其协同效应,也在光催化领域表现出巨大的优势。

1.4.2 ZnO/MOFs 复合材料

作为一种理想的光催化或光阳极材料,各种形态的 ZnO 已被设计和合成出来。然而,光载流子之间内部的快速复合大大限制了 ZnO 的催化效率。

2013年,Zhan等[70]首次采用简单的自模板法成功合成ZnO@ZIF-8纳米棒,如图1-10所示,在ZnO@ZIF-8纳米棒制备过程中,无须引入其他锌源,氧化锌纳米棒提供Zn^{2+},也作为ZIF-8形成的模板。更重要的是,得到的ZnO@ZIF-8纳米棒阵列具有良好的光电响应,同时,光生电子转移至电极基板,促使ZnO纳米棒光电流响应的显著增强。该研究为制备氧化锌基光催化剂开辟了一条新的途径。El-Hankari等[94]成功制备出核壳结构ZnO@ZIF-8,所制备的ZnO@ZIF-8有机物降解能力较ZnO有明显的提升。Ma等[95]采用一种原位自牺牲法制备了具有大量缺陷活性位点的ZnO@ZIF-8纳米复合材料。ZIF-8中非配位N原子与H_2O分子相互作用,形成氢键并形成质子通道,聚集大量氢离子,增加了催化剂与电解质的接触,抑制了光生载流子之间的内部复合,从而提高了光阳极的催化活性。

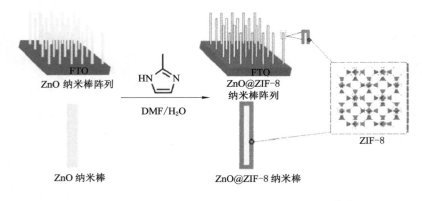

图1-10 自模板法合成ZnO@ZIF-8纳米棒示意图[70]

根据这一策略,多个研究小组成功合成了核壳结构的ZnO@MOFs,如Rad等[96]就成功制备了ZnO@MOF-46核壳异质结构,并研究了ZnO@MOF-46的溶剂型发光特性。结果表明,ZnO@MOF-46可以作为发光传感器检测小分子发光性能。Wang等[97]通过原位晶体生长法快速(60 min内)制得ZnO@ZIF-8核壳结构,该研究发现表面结合Zn^{2+}与2-甲基唑(Hmim)的速率与$Hmim/Zn^{2+}$的摩尔比相关,通过控制Hmim的浓度可以制备出理想的ZnO@ZIF-8核壳异质结构(图1-11),合成的ZnO@ZIF-8具有更薄的外壳(约30 nm)。Jia等[98]成功制备出1D ZnO@ZIF-8/67光电极,由于ZnO@ZIF-8/67复合材料能使电子快速地定向传输,从而提升了PEC性能,且在可见光照射下,ZnO@ZIF-8/67产生的光电流约等于ZnO的9.2

倍。研究表明,合成的 ZnO@ZIF-8 可以在 Cr(Ⅵ)和 MB 之间选择性地降低 Cr(Ⅵ)浓度,这是由于 ZIF-8 壳层的选择性渗透作用所导致的。ZnO/MOFs 复合材料的光催化性能较 ZnO 均具有显著提升,这是由于 MOFs 的导带产生的光生电子可以快速转移至 ZnO,同时 ZnO 的价带产生的光生空穴转移至 MOFs,抑制了光生电子和空穴的复合[98-100]。

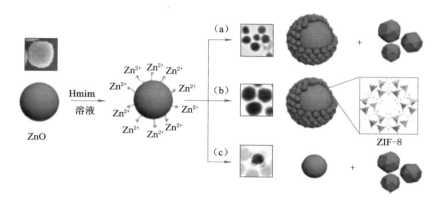

图 1-11　不同浓度的 Hmim 获得 ZnO@ZIF-8 结构示意图[97]

1.4.3　Fe_3O_4/MOFs 复合材料

在实际应用中,由于 MOFs 光催化剂具有高分散性,很难从反应溶液中分离出来进行回收[101-103]。解决这一问题,可用磁性粒子包覆 MOFs 光催化剂的方法进行改性。Zhang 等[104]制备新型多功能 Fe_3O_4@MIL-100(Fe)核壳纳米球,制备的 Fe_3O_4@MIL-100(Fe)对 MB 的降解表现出明显的光催化活性,高于一些典型的光催化剂,如 TiO_2 和 $g-C_3N_4$。Liu 等[105]通过逐层沉积和外延生长方法制得具有较大比表面积的新型 MOFs 杂化材料 Fe_3O_4@HKUST-1/MIL-100(Fe)(图 1-12),与纯 MOFs 相比,得到的杂化材料显著提高了比表面积,减小了内部孔隙,相同条件下,与纯 Fe_3O_4@MIL-100(Fe)相比,Fe_3O_4@HKUST-1/MIL-100(Fe)具有更高的 MB 去除率,这表明新形成的 Fe_3O_4@MOFs 复合材料保持了 MIL-100(Fe) 的催化性能。

Yue 等[106]通过水热法和过硫酸盐活化法成功制备高活性Fe_3O_4@MIL-101(Fe),Fe_3O_4@MIL-101(Fe)对偶氮染料 AO7 的去除率明显高于 Fe_3O_4 和 MIL-101(Fe),并探究了初始 pH 值、MIL-101 连接剂中氨基含量、MIL-101

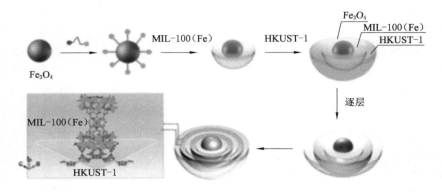

图 1-12　Fe$_3$O$_4$@HKUST-1/MIL-100(Fe)制备示意图[105]

中心金属离子的含量对 AO7 降解的影响。Sargazi 等[107]通过超声辅助反相胶束法制备出生物相容性好、可降解的 Ta-MOF@Fe$_3$O$_4$ 纳米材料。各种分析结果表明,该新型纳米材料具有较大的比表面积、不饱和金属中心和游离羧酸基团,这些特征使其成为一种新型的酶固定化候选材料。Huo 等[108]采用两步自组装方法成功制备 Fe$_3$O$_4$@UiO-66(图 1-13)。该方法是在 Fe$_3$O$_4$ 微球表面羧基辅助下直接外延生长 UiO-66,可以有效地避免额外的修饰过程,降低使用有毒有机试剂造成二次污染的风险,且合成的核壳复合材料具有比表面积高(124.8 m^2/g)和超顺磁性(磁化饱和度为 26.5 emu/g)等显著优点。

1.4.4　其他半导体/金属有机骨架复合材料

笔者以沸石咪唑类骨架材料 ZIF-8 为吸附材料、Zn$_2$GeO$_4$ 为半导体光催化剂,构建了新型 ZIF-8/Zn$_2$GeO$_4$ 复合光催化剂,借助 ZIFs 组元提高了催化剂的 CO$_2$ 吸附量,实现了液相体系催化剂表面 CO$_2$ 的富集和高效还原[22]。在可见光驱动的半导体光催化剂中,CdS 因具有优良的电子光学性能和合适的带隙(2.4 eV)而备受关注。福州大学王心晨教授等利用 Co-ZIF-9 与半导体光催化剂 C$_3$N$_4$ 和 CdS 复合,制备了 C$_3$N$_4$/Co-ZIF-9[109]和 CdS/Co-ZIF-9[110]复合光催化剂,复合后催化剂在可见光下将 CO$_2$ 还原成 CO 的活性得到了提高。He 等[111]采用合成后再处理法制备 CdS/MIL-101(Cr),首次证明了在 MOFs 上嵌入 CdS 可以显著提高 CdS 的光催化效率。Zeng 等[112]采用两步法制备 CdS@ZIF-8 纳米复合材料,制备过程中通过改变 Zn^{2+} 的浓度来制备多核和单核 CdS@ZIF-8 核壳结构,CdS@ZIF-8 甲酸光催化制 H$_2$ 性能均比

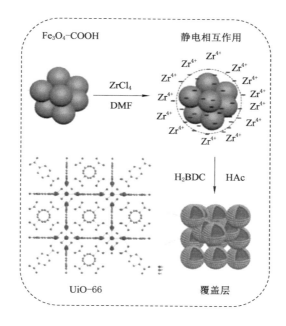

图 1-13 Fe$_3$O$_4$@UiO-66 制备示意图[108]

纯 CdS 高,而 CO 的生成率由 12.5% 降低到 5.7%。这是因为 ZIF-8 的孔径(3.4 Å)尺寸小于 CO 的直径(3.8 Å),生成的 CO 无法通过 ZIF-8 的孔隙。除此之外,相关 CdS/MOFs 研究还有 CdS/UiO-66[112]、CdS/MIL-53(Fe)[113]、CdS@MIL-125(Ti)[114]等。

  日本国立材料研究所叶金花教授等研究表明,UiO-66 与 C$_3$N$_4$ 光催化剂复合后,不仅大大提高了催化剂 CO$_2$ 的吸附能力,也促进了光生载流子的分离,从而提高了其光催化还原 CO$_2$ 的活性[115]。Argoub 等[116]采用溶胶-热合成法制备 MIL-101(Cr)@g-C$_3$N$_4$ 纳米复合材料,MIL-101(Cr)@g-C$_3$N$_4$ 纳米复合材料表面存在含有胺基的强碱性位点,使 MIL-101@g-C$_3$N$_4$ 纳米复合材料吸附 CO$_2$ 的能力较纯 MIL-101 有所提高。Gong 等[117]采用简便的水热法制备 g-C$_3$N$_4$/MIL-101(Fe)异质结构复合材料。该复合材料具有光吸附容量大、载流子分离效率高、表面活性位点丰富等优点,经过硫酸盐活化后的 g-C$_3$N$_4$/MIL-101(Fe)样品反应速率常数 k 达到 0.058 9 min^{-1},约为 g-C$_3$N$_4$ 的 8.9 倍。将 MIL-101(Fe)引入 g-C$_3$N$_4$ 中,不仅使形成的异质结具有合适的带隙对,有利于界面电荷载流子的分离,而且大大提高了(过硫酸盐)PS活化产生 SO_4^{2-} 的速率,因此对双酚A有更好的光催化分解能力,催化机理如图1-14所示。

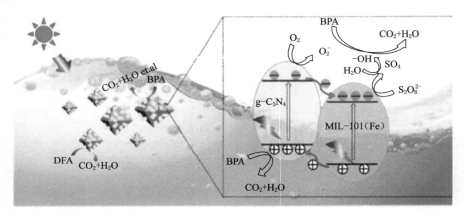

图1-14　g-C_3N_4/MIL-101(Fe)光催化机理图[117]

有学者采用溶剂热法制备了 MIL-101(Fe)/g-C_3N_4 复合材料,并将其用于光催化还原 Cr(Ⅵ),结果表明复合催化剂相比 MIL-101(Fe) 和 g-C_3N_4 具有更高的光催化性能[118]。孙为银等制备了 NH_2-MIL-101(Fe)/g-C_3N_4 Ⅱ型异质结,发现 NH_2-MIL-101(Fe) 和 g-C_3N_4 之间存在着有效的界面载流子传输,NH_2-MIL-101(Fe) 中的—NH_2 官能团促进了 CO_2 的吸附,从而增强了其 CO_2 的光催化还原活性[119]。

Abdelhameed 等[120]合成了 Ag_3PO_4 纳米颗粒和 Ag_3PO_4@NH_2-MIL-125 纳米复合材料,Ag_3PO_4@NH_2-MIL-125 纳米复合材料对 MB 和罗丹明 B 降解具有显著的光催化活性,分别是 P25 光催化剂的39倍和35倍。研究表明,Ag_3PO_4 纳米颗粒在降低 MIL-125-NH_2 的带隙方面发挥重要作用。蒋积菲[121]通过将 UiO-66 纳米球引入花状 $ZnIn_2S_4$ 微球中合成了一系列 $ZnIn_2S_4$/UiO-66 复合光催化剂,并探究其在可见光条件下的光催化析氢活性。优化后 $ZnIn_2S_4$/UiO-66 产氢速率是纯 $ZnIn_2S_4$ 产氢速率的3倍。该复合材料光催化活性提升的原因主要是由于 $ZnIn_2S_4$ 和 UiO-66 之间的快速电荷转移能力和有效的电子-空穴对分离效率。

为了克服 $BiVO_4$ 的电子-空穴分离和利用效率低的问题,Liu 等[122]通过水热法在 $BiVO_4$ 表面沉积一层超薄 MIL-101(Fe),合成了 MIL-101(Fe)/Mo:$BiVO_4$ 光阳极。优化后，MIL-101(Fe)/Mo:BiVO₄ 在 Na₂SO₄ 水溶液中 1.23 V(vs RHE)电压下的光电流密度为 4.01 mA/cm²，比原始 BiVO₄ 光阳极高 4 倍[图 1-15(a)]。此外，MIL-101(Fe)/Mo:BiVO₄ 光阳极具有良好的稳定性，光阳极性能的大幅度提高是由于超薄的 MIL-101(Fe)层作为析氧层助催化剂可以有效地扩展可见光响应区域，加速电荷分离，并提供更易接近的活性位点用于水氧化[图 1-15(b)]。

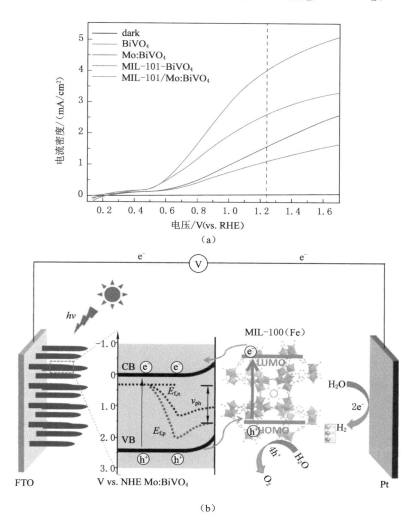

图 1-15　样品的 LSV 曲线及 MIL-101(Fe)/Mo:BiVO₄ 电荷转移示意图[122]

$Cd_{0.2}Zn_{0.8}S@UiO-66-NH_2$[123]、$CPO-27-Mg/TiO_2$[124]和$Pt/CeO_2@MOF$[125]等半导体/MOFs复合材料也表现出比复合前更好的光催化还原CO_2活性。这些相关研究说明半导体/MOFs复合材料的合成及其在多相光催化中的具有广泛应用,如光催化水分解、Cr(Ⅵ)还原、光催化CO_2还原、污染物降解和选择性有机转化等。虽然半导体/MOFs在光催化领域的应用还处于起步阶段,但实验结果表明,半导体与金属有机骨架材料复合形成新型光催化材料具有良好的应用前景。

1.5　MOFs衍生光催化材料

通过合理的结构和组分设计,MOFs材料可以表现出优良的光催化活性。但是其弱的热/化学稳定性以及弱的导电性抑制了MOFs光催化剂性能的提升。近年来研究发现MOFs可以作为多孔半导体材料,如金属氧化物、金属硫化物以及它们的异质结的前驱体。对这些前驱体材料进行化学处理或热处理得到MOFs衍生物,其形貌可控性强,继承了MOFs的多孔结构,有利于催化反应物的吸附,含有的金属氧化物位点可作为光活性中心,这些特性使得MOFs衍生材料在催化领域被广泛应用,如光催化中的CO_2还原、污染物降解等、电催化中的析氢反应(HER)、析氧反应(OER)、氧化还原反应(ORR)以及化学催化等方面。

MOFs衍生材料的形貌对性能的影响至关重要,其形貌主要取决于MOFs前驱体、热解温度和气体环境等。根据不同的应用要求,可以设计出多种用于光催化的衍生物,如碳基化合物、金属氧化物、金属硫化物以及由上述两种或三种材料组成的复合材料等。

1.5.1　MOFs衍生金属氧化物基光催化剂

Khaletskaya等[126]在溶剂热合成过程中,在NH_2-MIL-125表面沉积了预合成的金纳米颗粒,通过高温热解Au/NH_2-MIL-125复合材料制备了高光催化活性的Au/TiO_2纳米复合材料。以NH_2-MIL-125(Ti)为前驱体,经过溶剂化处理和高温热解可以实现MOFs结构的分解,得到多孔碳或介孔TiO_2[127]。Dekrafft等[128]以MIL-101纳米颗粒为前驱体,在其表面涂覆了一层无定型TiO_2,通过高温煅烧获得$Fe_2O_3@TiO_2$核壳结构(图1-16),负载贵金属Pt以后,$Fe_2O_3@TiO_2$核壳结构表现出优异的光催化水分解产氢活性。Zhan等[129]采用有序生长的MIL-68为前驱体,设计合成了$In_2S_3/CdIn_2S_4$

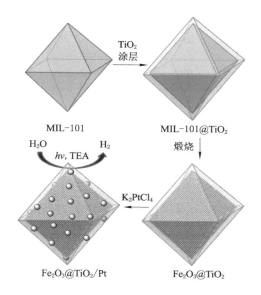

图 1-16　MOFs 衍生的 Fe_2O_3@TiO_2 结构示意图[128]

异质结纳米管(图 1-17)。In_2S_3 和 $CdIn_2S_4$ 之间形成纳米级界面接触，光生电荷的扩散距离大幅缩短，使得载流子能够快速迁移至表面参与反应。前驱体大的比表面积，使得生成的 In_2S_3/$CdIn_2S_4$ 比表面积大，CO_2 在催化剂表面吸附和富集，以及表面丰富的催化活性位点，这些因素使得 In_2S_3/$CdIn_2S_4$ 纳米管在可见光下能够有效光催化 CO_2 还原为 CO。以负载有 NiS 和 CdS 的 NH_2-MIL-125 为模板，Li 等[130] 热解得到中空的 CdS/TiO_2 催化剂，在催化剂相和助催化剂相之间原位形成的界面显著促进了光生电荷的分离和电子的迁移，表现出显著提升的光解水制氢性能。

金属有机骨架丰富的孔洞结构，能够为金属离子的掺杂提供空间，将掺杂金属离子或原子的 MOFs 进行煅烧可以得到多孔的掺杂材料，有利于扩展金属氧化物催化剂的种类。在 Fe_2O_3 纳米棒表面负载 NH_2-MIL-125(Ti)，经过煅烧得到 $Ti_xFe_{1-y}O_y$ 壳层，该掺杂氧化铁的光电流强度为纯 Fe_2O_3 纳米棒的 26.7 倍，实现了光催化性能的显著提升[131]。Chen 等[132] 在空气中煅烧金属有机骨架(Ti-MOF)以获得 Ti^{3+} 和氧空位掺杂的分布在碳基体中的锐钛矿和金红石异质结 TiO_2，其形貌结构如图 1-18 所示。结果表明，Ti^{3+} 和氧空位掺杂的异质结 TiO_2 表现出增强的可见光吸收和较高的载流子分离效率。最佳样品对四环素光催化降解效率达到 87.03%，金霉素的降解率达到 78.91%。

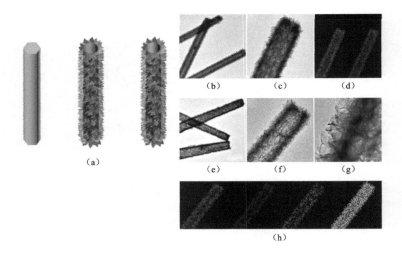

图 1-17　MOFs 衍生的 $In_2S_3/CdIn_2S_4$ 纳米管光催化剂示意图及
$In_2S_3/CdIn_2S_4$ 纳米管的结构表征[129]

此外,钛基 MOF 热解形成的碳层作为阻挡层可以防止 Ti^{3+} 和氧空位的氧化,使制备的材料具有良好的稳定性和重复性。

1.5.2　MOFs 衍生金属硫化物基光催化剂

二元或三元金属硫化物(如 CdS、ZnS、MoS_2、$ZnIn_2S_4$ 等)是一类重要的半导体光催化剂,在温和的条件下可以光催化促进一系列富有意义的氧化还原反应的发生。其带隙小,具有可见光响应,是太阳能驱动光催化反应非常有应用前景的光催化剂之一。但是金属硫化物光生电子和空穴容易复合以及其在水性介质中的高光腐蚀性限制了其实际应用。对此研究者们研究了各种方法来增强硫化物的光催化活性,包括增加比表面积和用助催化剂修饰硫化物或与其他材料形成固溶体等,这些方法均可以通过 MOF 衍生策略来实现。大量研究结果表明,使用 MOF 作为模板或前驱体,可以获得具有复杂中空纳米结构的硫化物,这种结构增强了催化剂的吸收能力,提高了光生电荷的分离效率并增加了比表面积,从而大大提高了金属硫化物的光催化活性。Xiao 等[133]利用热稳定的 MIL-53(Al)作为模板,得到了高比表面积的分级多孔 CdS(图 1-19)。研究结果表明,该结构可有效地抑制光生载流子的复合,与块状 CdS 相比,在可见光照射下多孔 CdS 光催化产氢活性得到了很大提升。Yun 等[134]基于普鲁士蓝前驱体合成了核壳结构的 CdS 材料,由于制备的

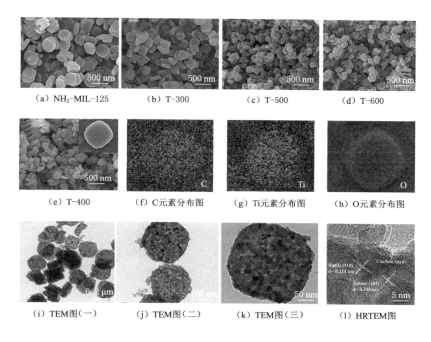

图 1-18 H$_2$-MIL-125、T-300、T-500、T-600 和 T-400 煅烧制备的
样品的 SEM 图以及 T-400 的 EDS 元素分布图、TEM 图、HRTEM 图[132]

CdS 特殊的三维结构、较高的比表面积及较小的纳米颗粒尺寸等，因而极大地提高了可见光照射下光催化水分解产氢速率[可达 3 051.4 μmol/(g·h)]。此外，由于 MOF 组成可变多孔，作为前驱体还可以与其他材料形成固溶体来修饰 CdS。Zhao 等[135]以 Ni、Zn 掺杂的 Cd-MOF 作为牺牲模板，通过溶剂热硫化和热退火制备了 NiS/Zn$_x$Cd$_{1-x}$S。通过调整 MOF 中的掺杂金属浓度，可以对异质结的化学组成和带隙进行微调，优化得到的 NiS/Zn$_{0.5}$Cd$_{0.5}$S 在可见光照射下的产氢活性可高达 16.78 mmol/(g·h)，并且具有很高的稳定性。

构建异质结结构是提高太阳能转换的光利用率和光生电荷分离与转移的有效途径。Pi 等[136]通过简单的 MOF 硫化过程衍生出 In$_2$S$_3$ 空心管异质结构，显著提高了其在可见光照射下的四环素光催化降解性能。可见光驱动降解效率的提高归因于多壁碳纳米管的电子受体（用于增强载流子分离）与具有较短载流子转移距离的 In$_2$S$_3$ 中空管中 MOF 的活性中心之间的协同作用，这不仅促进了载流子转移，而且有效抑制了复合速率，还可以改善可见光响应。

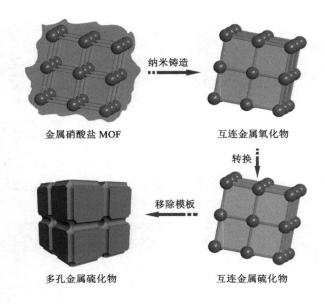

图 1-19 MOF 模板合成多孔金属氧化物/硫化物工艺示意图[133]

Lu 等[137]采用溶剂热法合成双金属 MOF(Cu-Zn-MOF),然后进行硫化处理转化得到多孔 CuS/ZnS 微球,显著提高了光的利用率,提供了丰富的暴露催化活性位点和有效的电子-空穴分离和传输,这使得 CuS/ZnS 微球具有优异的光催化分解水制氢活性。Zhang 等[138]采用 MOF 衍生获得空心六角棱镜结构的 In_2O_3,并与 $CdS-MoS_2$ 复合,得到 S 型异质结复合光催化剂,空心六角棱镜的纳米限制效应大大扩展了复合光催化剂的光谱吸收范围,极大地促进了电子-空穴对的分离,减弱表面氧化动力学,从而抑制了 CdS 表面光腐蚀反应,从而显著提高了半导体光催化剂的催化活性。

作为一种性能优异的可见光响应光催化剂,硫锌铟($ZnIn_2S_4$)由于具有无毒、合适的带隙、高物理化学稳定性和耐久性以及易于合成,引起了广泛的跨学科研究兴趣,成为光催化领域的一个新的研究热点。但 $ZnIn_2S_4$ 催化剂多为片状结构,结构调控比较困难。Fan 等[139]以有序大孔十四面体铈基金属有机骨架(Ce-MOFs)为模板和 Ce 离子源,通过一步水热法获得了具有空心纳米笼的 Ce 掺杂 $ZnIn_2S_4$ 光催化剂。具有空心纳米笼结构的 Ce 掺杂 $ZnIn_2S_4$ 继承了 Ce-MOF 模板的十四面体形状,且外壳由超薄纳米片组成。理论和实验结果表明,Ce 的掺杂和空心纳米笼的形成增加了光捕获和光生载流子的分离,从而提高了其光催化活性。Wu 等[140]采用 MOF 模板法制备了核壳结构的 CoP@ZnIn$_2$S$_4$ 光催化剂,其独特的组成和形态可以促进光激发载流子的分离,增强光吸收并提供高比表面积,CoP@ZnIn$_2$S$_4$ 光催化分解水 24 h 产 H$_2$ 为 0.103 mmol,且性能稳定,没有明显的光腐蚀产生。

1.5.3 MOFs 衍生碳基光催化剂

碳基材料由于导电性高、光谱吸收范围广,可以有效地收集太阳光,常与半导体光催化剂复合,获得高效复合光催化剂。碳本身可作为良好的电子传输层,也对光催化剂的表界面起到一定的修饰作用,自身良好的导电性也能形成电子通道,能有效地降低光生载流子的重组,从而使光电性能明显提升。MOFs 由金属中心和有机配体经过配位自组装而成,可采用高温热解方式转化成多孔碳材料和金属氧化物或硫化物等复合材料。相对于传统方法制备的光催化剂,MOFs 衍生制备的含碳复合光催化剂不但具有较大的比表面积、较多暴露的活性位点,而且碳结构也有利于光生电子传输,从而有效抑制光生电子-空穴的复合,大大提高光催化效率。MOFs 独特的结构特性使其成为较为理想的热解前驱体材料,进而设计合成多种含碳新型功能材料。从光催化角度而言,制备的含碳金属氧化物可将紫外光拓展到可见光范围,在一定程度上增强了太阳光的利用率。

Hussain 等[141]利用 MOF-5、ZIF-8 和 MOF-74M 衍生出三种 ZnO/C 复合物,并研究了其光催化活性,ZnO 均匀地分布在多孔碳中,能够吸附更多的亚甲基蓝分子(图 1-20)。而 ZnO 导带中的光生电子被转移到碳基质上,这促进了电荷分离,延长了光生载流子的寿命。MOF 前驱体决定了所得复合材料的晶体结构、掺杂分布、热稳定性和金属氧化物-碳重量比。MOF-5 衍生的多孔 ZnO/C 纳米复合材料不仅在可见光下表现出最高的光催化染料降解活性,而且在光催化水分解产氢方面比 MOF-74 和 ZIF-8 衍生的材料分别高出 9 倍和 4 倍。Wang 等[142]通过合理设计和控制 MOF 材料,制备了具有 3D 多孔结构、高比表面积的 g-C$_3$N$_4$/TiO$_2$/C 光催化剂,三维多孔结构提供了多维吸附富集位点,异质结促进了光生电子和空穴的分离,在光催化去除盐酸金霉素方面表现出高效的协同吸附光催化性能。笔者通过煅烧 MIL-101@TiO$_2$ 制备了 Cr$_2$O$_3$/C@TiO$_2$ 复合材料,通过 MIL-101 前驱体引入 Cr^{3+} 和 Cr$_2$O$_3$/C,提高了催化剂的可见光吸收能力,降低了光生电子和空穴的复合速率,从而提高了其光催化分解水产氢效率。Chen 等[143]采用 MIL-125(Ti)材料在适当温度下煅烧,成功制备了具有两种晶体类型(金红石和锐钛矿)和大比表面积的均匀片状 TiO$_2$/C 纳米复合材料。由于与 C 复合促进了光生电荷的分离和迁移，从而提高了其可见光光光催化活性。

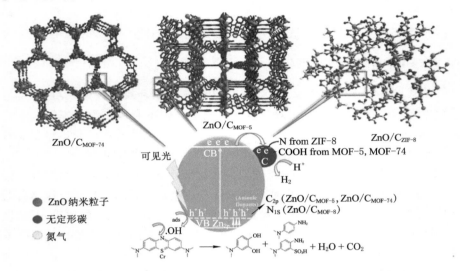

图 1-20　不同 MOF 前驱体制备的 ZnO/C 催化产氢和降解有机染料机制[141]

综上所述，MOF 是一类由金属离子/团簇和有机配体相连接组成的结晶性多孔材料，具有多样且可定制的结构以及高比表面积，广泛应用于气体吸附和分离、传感及催化等领域。MOF 由于具有众多独特优势而被用于光催化研究。首先，MOF 作为独特的材料，结合了均相催化剂精确的催化位点、高活性和选择性的优点以及非均相催化剂易于从反应产物中分离和可回收利用的优点；其次，MOF 孔隙率高和比表面积大有利于底物的吸附和富集，这有利于催化位点和反应物之间的接触和相互作用，因此能够提高催化效率；再次，MOF 结构可调使得可以通过多元方法或后合成修饰将所需的活性位点连接到金属节点或有机配体上，还可将其他活性物质封装到其孔道中，以协同增强催化作用；最后，金属有机骨架材料的多孔结构为电荷载流子创造了短的迁移路径，而且光敏剂或助催化剂可以灵活地引入骨架上或孔道中，进一步提高电子-空穴对的分离。此外，MOF 的大表面积、高孔隙率和可调的结构在 MOF 衍生的复合材料中得到了很好的保留，这使得 MOF 复合材料及 MOF 衍生的多孔材料成为多相催化领域研究的热点。

参考文献

[1] FUJISHIMA A, HONDA K. Electrochemical photolysis of water at a semiconductor electrode[J]. Nature, 1972, 238(5358): 37-38.

[2] CAREY J H, LAWRENCE J, TOSINE H M. Photodechlorination of PCB's in the presence of titanium dioxide in aqueous suspensions[J]. Bulletin of environmental contamination and toxicology, 1976, 16(6): 697-701.

[3] HALMANN M. Photoelectrochemical reduction of aqueous carbon dioxide on p-type gallium phosphide in liquid junction solar cells[J]. Nature, 1978, 275(5676): 115-116.

[4] DAI H, ZHOU Y, CHEN L, et al. Porous ZnO nanosheet arrays constructed on weaved metal wire for flexible dye-sensitized solar cells[J]. Nanoscale, 2013, 5(11): 5102-5108.

[5] DAI H, ZHOU Y, LIU Q, et al. Controllable growth of dendritic ZnO nanowire arrays on a stainless steel mesh towards the fabrication of large area, flexible dye-sensitized solar cells[J]. Nanoscale, 2012, 4(17): 5454-5460.

[6] TU W G, ZHOU Y, LIU Q, et al. An in situ simultaneous reduction-hydrolysis technique for fabrication of TiO_2-graphene 2D sandwich-like hybrid nanosheets: graphene-promoted selectivity of photocatalytic-driven hydrogenation and coupling of CO_2 into methane and ethane[J]. Advanced functional materials, 2013, 23(14): 1743-1749.

[7] WU H, LIU Q, ZHANG L, et al. Novel nanostructured WO_3@Prussian blue heterojunction photoanodes for efficient photoelectrochemical water splitting[J]. ACS applied energy materials, 2021, 4(11): 12508-12514.

[8] CHEN X Y, ZHOU Y, LIU Q, et al. Ultrathin, single-crystal WO_3 nanosheets by two-dimensional oriented attachment toward enhanced photocatalytic reduction of CO_2 into hydrocarbon fuels under visible light[J]. ACS applied materials and interfaces, 2012, 4(7): 3372-3377.

[9] FU J W, XU Q L, LOW J, et al. Ultrathin 2D/2D WO_3/g-C_3N_4 step-scheme H_2-production photocatalyst[J]. Applied catalysis B: environ-

mental,2019,243:556-565.

[10] CAO D P,LUO W J,FENG J Y,et al.Cathodic shift of onset potential for water oxidation on a Ti^{4+} doped Fe_2O_3 photoanode by suppressing the back reaction[J]. Energy and environmental science,2014,7(2):752-759.

[11] XU M,CHEN Y,MAO G B,et al.TiO_2 nanoparticle modified α-Fe_2O_3 nanospindles for improved photoelectrochemical water oxidation[J]. Materials express,2019,9(2):133-140.

[12] QIU T Y,WANG L,ZHOU B Y,et al.Molybdenum sulfide quantum dots decorated on TiO_2 for photocatalytic hydrogen evolution[J].ACS applied nano materials,2022,5(1):702-709.

[13] MAO G B,LI C X,LI Z D,et al.Efficient charge migration in TiO_2@PB nanorod arrays with core-shell structure for photoelectrochemical water splitting[J].CrystEngComm,2022,24(14):2567-2574.

[14] CHEN Y,MAO G B,TANG Y W,et al.Synthesis of core-shell nano-structured Cr_2O_3/C@TiO_2 for photocatalytic hydrogen production[J]. Chinese journal of catalysis,2021,42(1):225-234.

[15] XU M,WU H,TANG Y W,et al.One-step in situ synthesis of porous Fe^{3+}-doped TiO_2 octahedra toward visible-light photocatalytic conversion of CO_2 into solar fuel[J].Microporous and mesoporous materials,2020,309:110539.

[16] WEN J Q,XIE J,CHEN X B,et al.A review on g-C_3N_4-based photocatalysts[J].Applied surface science,2017,391:72-123.

[17] ONG W J,TAN L L,NG Y H,et al.Graphitic carbon nitride(g-C_3N_4)-based photocatalysts for artificial photosynthesis and environmental remediation: are we a step closer to achieving sustainability? [J]. Chemical reviews,2016,116(12):7159-7329.

[18] YAN S C,LI Z S,ZOU Z G.Photodegradation performance of g-C_3N_4 fabricated by directly heating melamine[J].Langmuir:the ACS journal of surfaces and colloids,2009,25(17):10397-10401.

[19] ZHENG Y,PAN Z M,WANG X C.Advances in photocatalysis in China [J].Chinese journal of catalysis,2013,34(3):524-535.

[20] LIU Q,WU D,ZHOU Y,et al.Single-crystalline, ultrathin $ZnGa_2O_4$

nanosheet scaffolds to promote photocatalytic activity in CO_2 reduction into methane[J]. ACS applied materials and interfaces, 2014, 6(4): 2356-2361.

[21] YAN S C, WANG J J, GAO H L, et al. An ion-exchange phase transformation to $ZnGa_2O_4$ nanocube towards efficient solar fuel synthesis[J]. Advanced functional materials, 2013, 23(6): 758-763.

[22] LIU Q, LUO Z X, LI L X, et al. ZIF-8/Zn_2GeO_4 nanorods with an enhanced CO_2 adsorption property in an aqueous medium for photocatalytic synthesis of liquid fuel[J]. Journal of materials chemistry A, 2013, 1(38): 11563.

[23] LIU Q, ZHOU Y, TU W G, et al. Solution-chemical route to generalized synthesis of metal germanate nanowires with room-temperature, light-driven hydrogenation activity of CO_2 into renewable hydrocarbon fuels[J]. Inorganic chemistry, 2014, 53(1): 359-364.

[24] LIU Q, ZHOU Y, KOU J H, et al. High-yield synthesis of ultralong and ultrathin Zn_2GeO_4 nanoribbons toward improved photocatalytic reduction of CO_2 into renewable hydrocarbon fuel[J]. Journal of the American chemical society, 2010, 132(41): 14385-14387.

[25] LIU Q, XU M, ZHOU B B, et al. Unique zinc germanium oxynitride hyperbranched nanostructures with enhanced visible-light photocatalytic activity for CO_2 reduction[J]. European journal of inorganic chemistry, 2017, 2017(15): 2195-2200.

[26] LIU Q, ZHOU Y, TIAN Z P, et al. Zn_2GeO_4 crystal splitting toward sheaf-like, hyperbranched nanostructures and photocatalytic reduction of CO_2 into CH_4 under visible light after nitridation[J]. Journal of materials chemistry, 2012, 22(5): 2033-2038.

[27] ZHANG W H, MOHAMED A R, ONG W J. Z-scheme photocatalytic systems for carbon dioxide reduction: where are we now? [J]. Angewandte chemie international edition, 2020, 59(51): 22894-22915.

[28] LINSEBIGLER A L, LU G Q, YATES J T. Photocatalysis on TiO_2 surfaces: principles, mechanisms, and selected results[J]. Chemical reviews, 1995, 95(3): 735-758.

[29] ZHOU Y, TIAN Z P, ZHAO Z Y, et al. High-yield synthesis of

ultrathin and uniform Bi_2WO_6 square nanoplates benefitting from photocatalytic reduction of CO_2 into renewable hydrocarbon fuel under visible light[J]. ACS applied materials and interfaces, 2011, 3(9): 3594-3601.

[30] LI H J, ZHOU Y, TU W G, et al. State-of-the-art progress in diverse heterostructured photocatalysts toward promoting photocatalytic performance[J]. Advanced functional materials, 2015, 25(7): 998-1013.

[31] WANG M, HAN Q T, LI L, et al. Construction of an all-solid-state artificial Z-scheme system consisting of Bi_2WO_6/Au/CdS nanostructure for photocatalytic CO_2 reduction into renewable hydrocarbon fuel[J]. Nanotechnology, 2017, 28(27): 274002.

[32] DONG J Q, HU J Q, LIU A Y, et al. Simple fabrication of Z-scheme $MgIn_2S_4$/Bi_2WO_6 hierarchical heterostructures for enhancing photocatalytic reduction of Cr(Ⅵ)[J]. Catalysis science and technology, 2021, 11(18): 6271-6280.

[33] LIU Q, ZHOU Y, MA Y, et al. Synthesis of highly crystalline $In_2Ge_2O_7$(En) hybrid sub-nanowires with ultraviolet photoluminescence emissions and their selective photocatalytic reduction of CO_2 into renewable fuel[J]. RSC advances, 2012, 2(8): 3247.

[34] LI Z D, ZHOU Y, BAO C X, et al. Vertically building Zn_2SnO_4 nanowire arrays on stainless steel mesh toward fabrication of large-area, flexible dye-sensitized solar cells[J]. Nanoscale, 2012, 4(11): 3490-3494.

[35] LI Z D, YANG H, ZHANG L F, et al. Stainless steel mesh-supported three-dimensional hierarchical SnO_2/Zn_2SnO_4 composite for the applications in solar cell, gas sensor, and photocatalysis[J]. Applied surface science, 2020, 502: 144113.

[36] LI P, ZHOU Y, TU W G, et al. Direct growth of $Fe_2V_4O_{13}$ nanoribbons on a stainless-steel mesh for visible-light photoreduction of CO_2 into renewable hydrocarbon fuel and degradation of gaseous isopropyl alcohol[J]. ChemPlusChem, 2013, 78(3): 274-278.

[37] FENG S C, CHEN X Y, ZHOU Y, et al. $Na_2V_4O_{16} \cdot xH_2O$ nanoribbons: large-scale synthesis and visible-light photocatalytic activity of CO_2 into

solar fuels[J]. Nanoscale, 2014, 6(3):1896-1900.

[38] KHAN S U M, AL-SHAHRY M, INGLER JR W B. Efficient photochemical water splitting by a chemically modified n-TiO_2[J]. Cheminform, 2003, 34(2):02016.

[39] FU W Z, ZHUANG P Y, OLIVERLAM CHEE M, et al. Oxygen vacancies in Ta_2O_5 nanorods for highly efficient electrocatalytic N_2 reduction to NH_3 under ambient conditions[J]. ACS sustainable chemistry and engineering, 2019, 7(10):9622-9628.

[40] SUZUKI T M, SAEKI S, SEKIZAWA K, et al. Photoelectrochemical hydrogen production by water splitting over dual-functionally modified oxide: p-Type N-doped Ta_2O_5 photocathode active under visible light irradiation[J]. Applied catalysis B: environmental, 2017, 202:597-604.

[41] LI Y, JIANG S H, XIAO J R, et al. Photocatalytic overall water splitting under visible light over an In-Ni-Ta-O-N solid solution without an additional cocatalyst[J]. International journal of hydrogen energy, 2014, 39(2):731-735.

[42] TSAI C W, CHEN H M, LIU R S, et al. Ni@NiO core-shell structure-modified nitrogen-doped $InTaO_4$ for solar-driven highly efficient CO_2 reduction to methanol[J]. The journal of physical chemistry C, 2011, 115(20):10180-10186.

[43] MAO G B, XU M, YAO S Y, et al. Direct growth of Cr-doped TiO_2 nanosheet arrays on stainless steel substrates with visible-light photoelectrochemical properties[J]. New journal of chemistry, 2018, 42(2):1309-1315.

[44] YUAN Y J, YU Z T, ZHANG J Y, et al. A copper(I) dye-sensitised TiO_2-based system for efficient light harvesting and photoconversion of CO_2 into hydrocarbon fuel[J]. Dalton transactions (Cambridge, England), 2012, 41(32):9594-9597.

[45] HAN Q T, LI L, GAO W, et al. Elegant construction of $ZnIn_2S_4$/$BiVO_4$ hierarchical heterostructures as direct Z-scheme photocatalysts for efficient CO_2 photoreduction[J]. ACS applied materials and interfaces, 2021, 13(13):15092-15100.

[46] KOCÍ K, OBALOVÁ L, MATĚJOVÁ L, et al. Effect of TiO_2 particle

size on the photocatalytic reduction of CO_2[J]. Applied catalysis B: environmental,2009,89(3/4):494-502.

[47] INOUE H,MATSUYAMA T,LIU B J,et al.Photocatalytic activities for carbon dioxide reduction of TiO_2 Microcrystals prepared in SiO_2 Matrices using a sol-gel method[J].Chemistry letters,1994,23(3):653-656.

[48] PAN J,WU X,WANG L Z,et al.Synthesis of anatase TiO_2 rods with dominant reactive {010} facets for the photoreduction of CO_2 to CH_4 and use in dye-sensitized solar cells[J].Chemical communications(Cambridge,England),2011,47(29):8361-8363.

[49] XIE Y P,LIU G,YIN L C,et al.Crystal facet-dependent photocatalytic oxidation and reduction reactivity of monoclinic WO_3 for solar energy conversion[J].Journal of materials chemistry,2012,22(14):6746.

[50] ZHANG N,OUYANG S,LI P,et al.Ion-exchange synthesis of a micro/mesoporous Zn_2GeO_4 photocatalyst at room temperature for photoreduction of CO_2[J].Chemical communications,2011,47(7):2041.

[51] MEI Z W,ZHANG N,OUYANG S X,et al.Photoassisted fabrication of zinc indium oxide/oxysulfide composite for enhanced photocatalytic H_2 evolution under visible-light irradiation[J]. Science and technology of advanced materials,2012,13(5):055001.

[52] SUN D R,YE L,LI Z H.Visible-light-assisted aerobic photocatalytic oxidation of amines to imines over NH_2-MIL-125(Ti)[J].Applied catalysis B:environmental,2015,164:428-432.

[53] SUN D R,FU Y H,LIU W J,et al.Studies on photocatalytic CO_2 reduction over NH_2-UiO-66(Zr) and its derivatives: towards a better understanding of photocatalysis on metal-organic frameworks[J].Chemistry-A European journal,2013,19(42):14279-14285.

[54] ISMAIL A A,BAHNEMANN D W,AL-SAYARI S A.Synthesis and photocatalytic properties of nanocrystalline Au,Pd and Pt photodeposited onto mesoporous RuO_2-TiO_2 nanocomposites[J].Applied catalysis A:general,2012(431/432):62-68.

[55] PAN P W,CHEN Y W.Photocatalytic reduction of carbon dioxide on $NiO/InTaO_4$ under visible light irradiation[J].Catalysis communica-

tions,2007,8(10):1546-1549.

[56] IIZUKA K,WATO T,MISEKI Y,et al.Photocatalytic reduction of carbon dioxide over Ag cocatalyst-loaded ALa$_4$Ti$_4$O$_{15}$(A=Ca,Sr,and Ba) using water as a reducing reagent[J].Journal of the American chemical society,2011,133(51):20863-20868.

[57] LIU X Q,IOCOZZIA J,WANG Y,et al.Noble metal-metal oxide nanohybrids with tailored nanostructures for efficient solar energy conversion,photocatalysis and environmental remediation[J].Energy and environmental science,2017,10(2):402-434.

[58] WANG P Q,BAI Y,LIU J Y,et al.One-pot synthesis of rutile TiO$_2$ nanoparticle modified anatase TiO$_2$ nanorods toward enhanced photocatalytic reduction of CO$_2$ into hydrocarbon fuels[J].Catalysis communications,2012,29:185-188.

[59] XU Q L,ZHANG L Y,CHENG B,et al.S-scheme heterojunction photocatalyst[J].Chem,2020,6(7):1543-1559.

[60] YU Z B,XIE Y P,LIU G,et al.Self-assembled CdS/Au/ZnO heterostructure induced by surface polar charges for efficient photocatalytic hydrogen evolution[J].Journal of materials chemistry A,2013,1(8):2773.

[61] TADA H,MITSUI T,KIYONAGA T,et al.All-solid-state Z-scheme in CdS-Au-TiO$_2$ three-component nanojunction system[J].Nature materials,2006,5(10):782-786.

[62] XING Z,SHEN S H,WANG M,et al.Efficient enhancement of solar-water-splitting by modified "Z-scheme" structural WO$_3^-$ W-Si photoelectrodes[J].Applied physics letters,2014,105(14):143902.

[63] ZHOU P,YU J G,JARONIEC M.All-solid-state Z-scheme photocatalytic systems[J].Advanced materials,2014,26(29):4920-4935.

[64] LI P,ZHOU Y,LI H J,et al.All-solid-state Z-scheme system arrays of Fe$_2$V$_4$O$_{13}$/RGO/CdS for visible light-driving photocatalytic CO$_2$ reduction into renewable hydrocarbon fuel[J].Chemical communications (Cambridge,England),2015,51(4):800-803.

[65] USMAN M,MENDIRATTA S,LU K L.Semiconductor metal-organic frameworks:future low-bandgap materials[J].Advanced materials,2017,29(6):1605071.

[66] KATAOKA Y, SATO K, MIYAZAKI Y, et al. Photocatalytic hydrogen production from water using porous material [Ru_2 (p-BDC)$_2$] N[J]. Energy and environmental science, 2009, 2(4): 397.

[67] FU Y H, SUN D R, CHEN Y J, et al. An amine-functionalized titanium metal-organic framework photocatalyst with visible-light-induced activity for CO_2 reduction[J]. Angewandte chemie, 2012, 124(14): 3420-3423.

[68] XU H Q, HU J H, WANG D K, et al. Visible-light photoreduction of CO_2 in a metal-organic framework: boosting electron-hole separation via electron trap states[J]. Journal of the American chemical society, 2015, 137(42): 13440-13443.

[69] LEE Y, KIM S, KANG J K, et al. Photocatalytic CO_2 reduction by a mixed metal(Zr/Ti), mixed ligand metal-organic framework under visible light irradiation [J]. Chemical communications (Cambridge, England), 2015, 51(26): 5735-5738.

[70] ZHAN W W, KUANG Q, ZHOU J Z, et al. Semiconductor@metal-organic framework core-shell heterostructures: a case of ZnO@ZIF-8 nanorods with selective photoelectrochemical response[J]. Journal of the American chemical society, 2013, 135(5): 1926-1933.

[71] WANG D K, HUANG R K, LIU W J, et al. Fe-based MOFs for photocatalytic CO_2 reduction: role of coordination unsaturated sites and dual excitation pathways[J]. ACS catalysis, 2014, 4(12): 4254-4260.

[72] WANG J, ZHANG J, PEH S B, et al. Dimensional impact of metal-organic frameworks in catalyzing photoinduced hydrogen evolution and cyanosilylation reactions[J]. ACS applied energy materials, 2019, 2(1): 298-304.

[73] ROSEN B A, SALEHI-KHOJIN A, THORSON MR, et al. Ionic liquid-mediated selective conversion of CO_2 to CO at low overpotentials[J]. Science, 2011, 334(6056): 643-644.

[74] ZHAO M T, YUAN K, WANG Y, et al. Metal-organic frameworks as selectivity regulators for hydrogenation reactions[J]. Nature, 2016, 539(7627): 76-80.

[75] WANG S B, YAO W S, LIN J L, et al. Cobalt imidazolate metal-organic frameworks photosplit CO_2 under mild reaction conditions[J]. Ange-

wandte chemie international edition,2014,53(4):1034-1038.

[76] WANG H,YUAN X Z,WU Y,et al.Synthesis and applications of novel graphitic carbon nitride/metal-organic frameworks mesoporous photocatalyst for dyes removal[J].Applied catalysis B:environmental,2015,174/175:445-454.

[77] WANG H,YUAN X Z,WU Y,et al.In situ synthesis of In_2S_3@MIL-125(Ti) core-shell microparticle for the removal of tetracycline from wastewater by integrated adsorption and visible-light-driven photocatalysis[J]. Applied catalysis B:environmental,2016,186:19-29.

[78] HORIUCHI Y,TOYAO T,SAITO M,et al.Visible-light-promoted photocatalytic hydrogen production by using an amino-functionalized Ti(Ⅳ) metal-organic framework[J].The journal of physical chemistry C, 2012,116(39):20848-20853.

[79] XIAO J D,SHANG Q C,XIONG Y J,et al.Boosting photocatalytic hydrogen production of a metal-organic framework decorated with platinum nanoparticles:the platinum location matters[J].Angewandte chemie international edition,2016,55(32):9389-9393.

[80] XIAO J D,HAN L L,LUO J,et al.Integration of plasmonic effects and Schottky junctions into metal-organic framework composites:steering charge flow for enhanced visible-light photocatalysis[J].Angewandte chemie international edition,2018,57(4):1103-1107.

[81] DOZZI M,SELLI E.Specific facets-dominated anatase TiO_2:fluorine-mediated synthesis and photoactivity[J].Catalysts,2013,3(2):455-485.

[82] KARTHIKEYAN C,ARUNACHALAM P,RAMACHANDRAN K,et al.Recent advances in semiconductor metal oxides with enhanced methods for solar photocatalytic applications[J].Journal of alloys and compounds,2020,828:154281.

[83] ZHANG Y Z,XIA B Q,RAN J R,et al.Atomic-level reactive sites for semiconductor-based photocatalytic CO_2 reduction[J].Advanced energy materials,2020,10(9):1903879.

[84] WANG C C,WANG X,LIU W.The synthesis strategies and photocatalytic performances of TiO_2/MOFs composites:a state-of-the-art review [J].Chemical engineering journal,2020,391:123601.

[85] LI R, HU J H, DENG M S, et al. Metal-organic frameworks: integration of an inorganic semiconductor with a metal-organic framework: a platform for enhanced gaseous photocatalytic reactions[J]. Advanced materials, 2014, 26(28): 4907.

[86] CHANG N, HE D Y, LI Y X, et al. Fabrication of TiO_2@MIL-53 core-shell composite for exceptionally enhanced adsorption and degradation of nonionic organics[J]. RSC advances, 2016, 6(75): 71481-71484.

[87] LIU Q, ZHOU B B, XU M, et al. Integration of nanosized ZIF-8 particles onto mesoporous TiO_2 nanobeads for enhanced photocatalytic activity[J]. RSC advances, 2017, 7(13): 8004-8010.

[88] ZENG X, HUANG L Q, WANG C N, et al. Sonocrystallization of ZIF-8 on electrostatic spinning TiO_2 nanofibers surface with enhanced photocatalysis property through synergistic effect[J]. ACS applied materials and interfaces, 2016, 8(31): 20274-20282.

[89] WANG M T, WANG D K, LI Z H. Self-assembly of CPO-27-Mg/TiO_2 nanocomposite with enhanced performance for photocatalytic CO_2 reduction[J]. Applied catalysis B: environmental, 2016, 183: 47-52.

[90] SHENG H B, CHEN D Y, LI N J, et al. Urchin-inspired TiO_2@MIL-101 double-shell hollow particles: adsorption and highly efficient photocatalytic degradation of hydrogen sulfide[J]. Chemistry of materials, 2017, 29(13): 5612-5616.

[91] MÜLLER M, ZHANG X, WANG Y, et al. Nanometer-sized titania hosted inside MOF-5[J]. Chemical communications (Cambridge, England), 2009(1): 119-121.

[92] WANG H M, YU T, TAN X, et al. Enhanced photocatalytic oxidation of isopropanol by HKUST-1@TiO_2 core-shell structure with ultrathin anatase porous shell: toxic intermediate control[J]. Industrial and engineering chemistry research, 2016, 55(29): 8096-8103.

[93] 李一新. 金属有机骨架、二氧化钛复合材料的制备及其在有机物处理中的应用[D]. 天津: 天津工业大学, 2017.

[94] EL-HANKARI S, AGUILERA-SIGALAT J, BRADSHAW D. Surfactant-assisted ZnO processing as a versatile route to ZIF composites and hollow architectures with enhanced dye adsorption[J]. Journal of materials

chemistry A,2016,4(35):13509-13518.
[95] MA Y L,WANG X,SUN X D,et al.Self-sacrificed construction of defect-rich ZnO@ZIF-8 nanocomposites with enhanced photocurrent properties[J].Inorganic chemistry frontiers,2020,7(4):1046-1053.
[96] RAD M,DEHGHANPOUR S.ZnO as an efficient nucleating agent and morphology template for rapid,facile and scalable synthesis of MOF-46 and ZnO@MOF-46 with selective sensing properties and enhanced photocatalytic ability[J].RSC advances,2016,6(66):61784-61793.
[97] WANG X B,LIU J,LEONG S,et al.Rapid construction of ZnO@ZIF-8 heterostructures with size-selective photocatalysis properties[J].ACS applied materials and interfaces,2016,8(14):9080-9087.
[98] JIA G R,LIU L L,ZHANG L,et al.1D alignment of ZnO@ZIF-8/67 nanorod arrays for visible-light-driven photoelectrochemical water splitting[J].Applied surface science,2018,448:254-260.
[99] 李志猛.含氮类配体MOFs和中空ZnO@ZIF-8复合材料的合成及其性质研究[D].海口:海南大学,2017.
[100] HUANG C W,NGUYEN V H,ZHOU S R,et al.Metal-organic frameworks:preparation and applications in highly efficient heterogeneous photocatalysis[J].Sustainable energy and fuels,2020,4(2):504-521.
[101] ZHANG T,ZHANG X F,YAN X J,et al.Synthesis of Fe_3O_4@ZIF-8 magnetic core-shell microspheres and their potential application in a capillary microreactor[J].Chemical engineering journal,2013,228:398-404.
[102] MIN X,YANG W T,HUI Y F,et al.Fe_3O_4@ZIF-8:a magnetic nanocomposite for highly efficient UO_2^{2+} adsorption and selective UO_2^{2+}/Ln^{3+} separation[J].Chemical communications (Cambridge, England),2017,53(30):4199-4202.
[103] ZOU Y L,ZHANG Y T,LIU X Y,et al.Solvent-free synthetic Fe_3O_4@ZIF-8 coated lipase as a magnetic-responsive Pickering emulsifier for interfacial biocatalysis[J].Catalysis letters,2020,150(12):3608-3616.
[104] ZHANG C F,QIU L G,KE F,et al.A novel magnetic recyclable photocatalyst based on a core-shell metal-organic framework Fe_3O_4@MIL-100(Fe)

for the decolorization of methylene blue[J].Journal of materials chemistry A,2013,1(45):14329.

[105] LIU J,YANG F,ZHANG Q L,et al.Construction of hierarchical Fe_3O_4@HKUST-1/MIL-100(Fe) microparticles with large surface area through layer-by-layer deposition and epitaxial growth methods [J].Inorganic chemistry,2019,58(6):3564-3568.

[106] YUE X X,GUO W L,LI X H,et al.Core-shell Fe_3O_4@MIL-101(Fe) composites as heterogeneous catalysts of persulfate activation for the removal of Acid Orange 7[J].Environmental science and pollution research international,2016,23(15):15218-15226.

[107] SARGAZI G,AFZALI D,EBRAHIMI A K,et al.Ultrasound assisted reverse micelle efficient synthesis of new Ta-MOF@Fe_3O_4 core/shell nanostructures as a novel candidate for lipase immobilization[J].Materials science and engineering:C,2018,93:768-775.

[108] HUO J B,XU L,CHEN X X,et al.Direct epitaxial synthesis of magnetic Fe_3O_4@UiO-66 composite for efficient removal of arsenate from water[J].Microporous and mesoporous materials,2019,276:68-75.

[109] WANG S B,LIN J L,WANG X C.Semiconductor-redox catalysis promoted by metal-organic frameworks for CO_2 reduction[J].Physical chemistry chemical physics:PCCP,2014,16(28):14656-14660.

[110] WANG S B,WANG X C.Photocatalytic CO_2 reduction by CdS promoted with a zeolitic imidazolate framework[J].Applied catalysis B:environmental,2015,162:494-500.

[111] HE J,YAN Z Y,WANG J Q,et al.Significantly enhanced photocatalytic hydrogen evolution under visible light over CdS embedded on metal-organic frameworks[J].Chemical communications(Cambridge,England),2013,49(60):6761-6763.

[112] ZENG M,CHAI Z G,DENG X,et al.Core-shell CdS@ZIF-8 structures for improved selectivity in photocatalytic H_2 generation from formic acid[J].Nano research,2016,9(9):2729-2734.

[113] CHATURVEDI G,KAUR A,KANSAL S K.CdS-decorated MIL-53(Fe) microrods with enhanced visible light photocatalytic performance for

the degradation of ketorolac tromethamine and mechanism insight [J]. The journal of physical chemistry C,2019,123(27):16857-16867.

[114] RAHMANI A,EMROOZ H B M,ABEDI S,et al. Synthesis and characterization of CdS/MIL-125(Ti) as a photocatalyst for water splitting[J]. Materials science in semiconductor processing,2018,80:44-51.

[115] SHI L,WANG T,ZHANG H B,et al. Electrostatic self-assembly of nanosized carbon nitride nanosheet onto a zirconium metal-organic framework for enhanced photocatalytic CO_2 reduction[J]. Advanced functional materials,2015,25(33):5360-5367.

[116] ARGOUB A,GHEZINI R,BACHIR C,et al. Synthesis of MIL-101@g-C_3N_4 nanocomposite for enhanced adsorption capacity towards CO_2 [J]. Journal of porous materials,2018,25(1):199-205.

[117] GONG Y,YANG B,ZHANG H,et al. A g-C_3N_4/MIL-101(Fe) heterostructure composite for highly efficient BPA degradation with persulfate under visible light irradiation[J]. Journal of materials chemistry A,2018,6(46):23703-23711.

[118] ZHAO F P,LIU Y P,HAMMOUDA S B,et al. MIL-101(Fe)/g-C_3N_4 for enhanced visible-light-driven photocatalysis toward simultaneous reduction of Cr(Ⅵ) and oxidation of bisphenol A in aqueous media [J]. Applied catalysis B:environmental,2020,272:119033.

[119] DAO X Y,XIE X F,GUO J H,et al. Boosting photocatalytic CO_2 reduction efficiency by heterostructures of NH_2-MIL-101(Fe)/g-C_3N_4 [J]. ACS applied energy materials,2020,3(4):3946-3954.

[120] ABDELHAMEED R M,TOBALDI D M,KARMAOUI M. Engineering highly effective and stable nanocomposite photocatalyst based on NH_2-MIL-125 encirclement with Ag_3PO_4 nanoparticles[J]. Journal of photochemistry and photobiology A:chemistry,2018,351:50-58.

[121] 蒋积菲. 金属有机骨架/半导体复合材料的制备及光催化产氢性能研究[D]. 南京:东南大学,2019.

[122] LIU C H,LUO H,XU Y,et al. Synergistic cocatalytic effect of ultrathin metal-organic framework and Mo-dopant for efficient photoelectrochemical water oxidation on $BiVO_4$ photoanode[J]. Chemical engineering journal,2020,384:123333.

[123] WANG M T, WANG D K, LI Z H. Self-assembly of CPO-27-Mg/TiO_2 nanocomposite with enhanced performance for photocatalytic CO_2 reduction[J]. Applied catalysis B: environmental, 2016, 183: 47-52.

[124] SU Y, ZHANG Z, LIU H, et al. $Cd_{0.2}Zn_{0.8}$S@UiO-66-NH_2 nanocomposites as efficient and stable visible-light-driven photocatalyst for H_2 evolution and CO_2 reduction[J]. Applied catalysis B: environmental, 2017, 200: 448-457.

[125] ZHAN W W, KUANG Q, ZHOU J Z, et al. Semiconductor@metal-organic framework core-shell heterostructures: a case of ZnO@ZIF-8 nanorods with selective photoelectrochemical response[J]. Journal of the American chemical society, 2013, 135(5): 1926-1933.

[126] KHALETSKAYA K, POUGIN A, MEDISHETTY R, et al. Fabrication of gold/titania photocatalyst for CO_2 reduction based on pyrolytic conversion of the metal-organic framework NH_2-MIL-125(Ti) loaded with gold nanoparticles[J]. Chemistry of materials, 2015, 27(21): 7248-7257.

[127] GU Z Z, CHEN L Y, LI X Z, et al. NH_2-MIL-125(Ti)-derived porous cages of titanium oxides to support Pt-Co alloys for chemoselective hydrogenation reactions[J]. Chemical science, 2018, 10(7): 2111-2117.

[128] DEKRAFFT K E, WANG C, LIN W B. Metal-organic framework templated synthesis of Fe_2O_3/TiO_2 nanocomposite for hydrogen production[J]. Advanced materials, 2012, 24(15): 2014-2018.

[129] ZHAN W W, SUN L M, HAN X G. Recent progress on engineering highly efficient porous semiconductor photocatalysts derived from metal-organic frameworks[J]. Nano-micro letters, 2019, 11(1): 1.

[130] LI N X, HUANG H L, BIBI R, et al. Noble-metal-free MOF derived hollow CdS/TiO_2 decorated with NiS cocatalyst for efficient photocatalytic hydrogen evolution[J]. Applied surface science, 2019, 476: 378-386.

[131] LI C H, HUANG C L, CHUAH X F, et al. Ti-MOF derived $Ti_xFe_{1-x}O_y$ shells boost Fe_2O_3 nanorod cores for enhanced photoelectrochemical water oxidation[J]. Chemical engineering journal, 2019, 361: 660-670.

[132] CHEN X Y, PENG X, JIANG L B, et al. Photocatalytic removal of an-

tibiotics by MOF-derived Ti^{3+}-and oxygen vacancy-doped anatase/rutile TiO_2 distributed in a carbon matrix[J]. Chemical engineering journal,2022,427:130945.

[133] XIAO J D,JIANG H L. Thermally stable metal-organic framework-templated synthesis of hierarchically porous metal sulfides: enhanced photocatalytic hydrogen production[J]. Small,2017,13(28):1700632.

[134] YUN S,DAN A,HONG L,et al. MOF-derived yolk-shell CdS microcubes with enhanced visible-light photocatalytic activity and stability for hydrogen evolution[J]. Journal of materials chemistry A,2017,5(18):8680-8689.

[135] ZHAO X X,FENG J R,LIU J,et al. An efficient, visible-light-driven, hydrogen evolution catalyst $NiS/Zn_xCd_{1-x}S$ nanocrystal derived from a metal-organic framework[J]. Angewandte chemie international edition,2018,57(31):9790-9794.

[136] PI Y H,JIN S,LI X Y,et al. Encapsulated MWCNT@MOF-derived In_2S_3 tubular heterostructures for boosted visible-light-driven degradation of tetracycline[J]. Applied catalysis B:environmental,2019,256:117882.

[137] LU J Q,ZHANG J,CHEN Q,et al. Porous CuS/ZnS microspheres derived from a bimetallic metal-organic framework as efficient photocatalysts for H_2 production[J]. Journal of photochemistry and photobiology A:chemistry,2019,380:111853.

[138] ZHANG L J,JIANG X D,JIN Z L,et al. Spatially separated catalytic sites supplied with the $CdS-MoS_2-In_2O_3$ ternary dumbbell S-scheme heterojunction for enhanced photocatalytic hydrogen production[J]. Journal of materials chemistry A,2022,10(19):10715-10728.

[139] FAN H T,JIN Y J,LIU K C,et al. One-step MOF-templated strategy to fabrication of Ce-doped $ZnIn_2S_4$ tetrakaidecahedron hollow nanocages as an efficient photocatalyst for hydrogen evolution[J]. Advanced science,2022,9(9):2104579.

[140] WU B Y,LIU N,LU L L,et al. A MOF-derived hierarchical CoP@$ZnIn_2S_4$ photocatalyst for visible light-driven hydrogen evolution[J]. Chemical communications,2022,58(46):6622-6625.

[141] HUSSAIN M Z, PAWAR G S, HUANG Z, et al. Porous ZnO/Carbon nanocomposites derived from metal organic frameworks for highly efficient photocatalytic applications: a correlational study[J]. Carbon, 2019, 146: 348-363.

[142] WANG Y R, WANG A N, PAN J, et al. Metal-organic complex-derived 3D porous carbon-supported g-C_3N_4/TiO_2 as photocatalysts for the efficient degradation of antibiotic[J]. CrystEngComm, 2021, 23(26): 4717-4723.

[143] CHEN J F, ZHANG X D, BI F K, et al. A facile synthesis for uniform tablet-like TiO_2/C derived from Materials of Institut Lavoisier-125(Ti) [MIL-125(Ti)] and their enhanced visible light-driven photodegradation of tetracycline[J]. Journal of colloid and interface science, 2020, 571: 275-284.

第 2 章　TiO_2/ZIF-8 复合材料的制备及其光催化性能研究

2.1　引言

六价铬 Cr(Ⅵ)由于具有高毒性和极好的溶解性,是工业废水中最有毒和最具危险的污染物之一[1-3],与 Cr(Ⅵ)相比,Cr(Ⅲ)毒性较小,易于在中性或碱性溶液中沉淀,因此废水中 Cr(Ⅵ)的优选处理方法是先把 Cr(Ⅵ)还原为 Cr(Ⅲ),随后从水溶液中吸附 Cr(Ⅲ)[4]。在不同的处理方法中(如膜分离、化学沉淀、离子交换、吸附和光催化等),光催化被证明是将 Cr(Ⅵ)还原为 Cr(Ⅲ)的有效途径之一[5-7]。迄今为止,大多数报道的光催化剂都是半导体材料,如 TiO_2[5-7]、CdS[8]、SnS_2[9]、Ag_2S[10]等,光催化过程中反应物的吸附或渗透对量子效率起着重要的作用[11-13]。通过增加光催化剂的比表面积来提供更多的活性位点和提升吸附性能[11],已被证明是提升光催化活性的有效途径。然而,传统半导体材料的比表面积有限,限制了它们的光催化活性。

金属有机骨架材料(MOFs)由于具有高的比表面积、大的孔径、可调节和良好的纳米孔道结构以及化学可裁剪性,在分子识别、气体分离、催化和药物输送等方面有着广泛的应用[14]。而且,在水分解[15-16]、CO_2 还原[17-19]、有机污染物光降解[20-22]、醇的光催化氧化[23]和光催化还原 Cr(Ⅵ)[24-26]等领域有着广泛的应用。例如,Li 等[21]合成了一种氨基功能化的 NH_2-MIL-125(Ti)光催化剂,以乙腈-三乙醇胺为牺牲剂,在可见光下催化 CO_2 还原。Mahata 等[27]分别使用 Ni、Co 和 Zn 基 MOFs 作为光催化剂降解有机染料。此外,Fe(Ⅲ)基 MOFs 在可见光下对 Cr(Ⅵ)还原具有良好的光催化活性[24,26]。尽管许多类型的 MOFs 被证明是新型光催化剂,但到目前为止,与传统的半导体光催化剂相比，MOFs 光催化剂的光催化效率依然很低，这是由于其在激发子产生和电荷分离方面的低效率，这限制了其在实际中的应用。另外，在反应过程中 MOFs 的稳定性是一个需要进一步研究的重要问题。

将高比表面积和吸附性能的 MOFs 与高活性的半导体光催化剂结合起来，为新型光催化剂的制备提供了一种很有前途的策略。近年来，纳米结构半导体材料/MOFs 形成复合材料作为光催化剂得到了广泛的研究，如 C_3N_4/Co-ZIF-9[28]、$Cd_{0.2}Zn_{0.8}S$@UiO-66-NH_2[29]、TiO_2@MIL-53[30]、TiO_2/ZIF-8[31]、$Cu_3(BTC)_2$@TiO_2[32]、ZnO@ZIF-8[33-34]、Fe_2O_3@MIL-101[35]、UiO-66/g-C_3N_4[36]、ZIF-8/Zn_2GeO_4[37]、CPO-27-Mg/TiO_2[38]等。研究表明，半导体与 MOFs 复合后，由于其协同效应而表现出超过单个组分的性能，除了增加复合材料的比表面积外，半导体与 MOF 之间也会发生电荷转移，从而大大抑制了半导体/MOFs 复合材料中的电子-空穴复合，提供了长寿命的电子。尽管人们对半导体纳米结构/MOFs 复合材料的研究越来越多，但对这些基于 MOFs 的杂化光催化剂的研究仍然在起步阶段。

本书设计与制备了 TiO_2/ZIF-8 复合光催化剂，选择金属有机骨架材料中的 ZIF-8 是由于其具有大的比表面积、良好的热稳定性和化学稳定性[39]。用 ZIF-8 纳米颗粒修饰合成的介孔 TiO_2 球，制备了新型 TiO_2/ZIF-8 复合纳米粒子，并对 TiO_2/ZIF-8 纳米粒子的形貌、吸附性能、比表面积和光催化还原 Cr(Ⅵ)性能进行表征。与相同尺寸的 TiO_2 球相比，在室温、300 mW/cm^2 光照射下，TiO_2/ZIF-8 将 Cr(Ⅵ)还原为 Cr(Ⅲ)的催化活性明显增强。通过对周期性照射的瞬态光电流响应和合成物的多次光催化还原实验，考察了新催化剂的稳定性和长期性能利用回收光催化剂的 Cr(Ⅳ)废液。据我们所知，TiO_2/ZIF-8 纳米复合材料的设计合成及其在 Cr(Ⅳ)光还原中的应用，能够激发人们对利用 MOFs 制造其他高性能半导体/MOFs 复合材料的兴趣。

2.2 实验部分

2.2.1 介孔 TiO_2 制备

通过改进已报道方法制备介孔 TiO_2 球[40]。取 0.2 mL 钛酸四丁酯(TBT)加入 10 mL 乙二醇溶液中，在室温下搅拌 24 h，获得透明溶液，将上述溶液倒入 100 mL 丙酮溶液中(含水约 0.3%)搅拌 15 min，然后静置 1 h。白色沉淀物用离心方法分离，用乙醇和去离子水清洗几次，以去除颗粒表面的多余乙二醇。取上述白色粉末加入含 10 mL 去离子水的聚四氟乙烯内衬 25 mL 的不锈钢反应釜,随后放入烘箱,设置烘箱温度为 180 ℃,保温时间为 24 h,然后自然冷却到室温。反应产物用酒精彻底清洗几次,最后用乙醇分散后,将分散液放入干燥箱 60 ℃、12 h。

2.2.2 ZIF-8 制备

ZIF-8 根据之前报道的方法[37]制备。取 0.589 g $Zn(NO_3)_2 \cdot 6H_2O$、1.298 g 2-甲基咪唑分别溶于 40 mL 甲醇溶液中,随后将两种溶液混合,室温条件下搅拌 2 h。反应产物用去离子水和乙醇分别离心洗涤 3 次,最后用乙醇分散后,将分散液放入干燥箱 60 ℃、12 h。

2.2.3 TiO_2/ZIF-8 复合材料制备

取 0.4 g 制备的介孔 TiO_2,加入溶液(1)0.589 g $Zn(NO_3)_2 \cdot 6H_2O$ 的 40 mL 甲醇溶液的中,搅拌 1 h;溶液(2)1.298 g 2-甲基咪唑溶于 40 mL 甲醇溶液;在搅拌的条件下,将溶液(2)缓慢倒入溶液(1)中,再搅拌 1 h(ZIF-8 与 TiO_2 的理论质量比为 1∶2)。反应产物用去离子水和乙醇分别离心洗涤 3 次,最后用乙醇分散后,将分散液放入干燥箱 60 ℃、12 h。

2.2.4 样品表征手段

利用日本电子 JEM-2100 透射电子显微镜和日立 S-4800 扫描电子显微镜对样品的微观结构进行了形貌分析。利用德国布鲁 D8 型衍射仪对样品的物相结构进行了表征。利用日本岛津紫外可见分光光度计 UV-3600,在 300~800 nm 范围内分析了样品的光吸收特征和带隙结构。利用美国麦克公司 ASAP 2020 比表面积及孔径分析仪测定样品的 N_2 吸附-脱附曲线,比表面积由测得曲线的线性部分计算获得。利用日本岛津公司 DSC-60A 自动差热热重仪同时测定装置对样品的稳定性和含量进行测定。

2.2.5 光电化学测量

光电极制备:在 1 mL 乙醇中加入 10 mg 制备的样品,超声处理 1 h,使样品在乙醇溶液中均匀分散。取 10 μL 分散液滴在 FTO 玻璃基片上(暴露面积约 1.0 cm^2),随后在 60 ℃真空干燥箱内干燥 1 h。重复上述步骤 5 次,获得光催化剂在 FTO 玻璃基片上均匀覆盖的光电极。

光电性能测试:本测试使用电化学工作站(上海振华 CHI630D),测试使用三电极体系,以制备的光电极为工作电极,铂丝电极为对电极,Ag/AgCl电极为参比电极,电解液为 0.1 mol/L Na$_2$SO$_4$ 水溶液(pH=7)电极在 300 W 氙灯辐照条件下,进行循环伏安曲线测试,扫描速率为 30 mV/s。

2.2.6 光催化还原 Cr(Ⅵ)性能测试

实验制得催化剂的光催化活性是在 300 W 氙灯光照下,用 Cr(Ⅵ)水溶液中 Cr(Ⅵ)离子去除率来表征。选择重铬酸钾(K$_2$Cr$_2$O$_7$)作为 Cr(Ⅵ)源。室温下,在石英反应器加入 20 mg 所制备的催化剂和 40 mL 的 Cr(Ⅵ)溶液[Cr(Ⅵ)离子浓度为 20 mg/L,pH=7]的,然后通过高纯氮气以排除溶液中溶入的氧气。最后,加入 5 mg 的草酸铵(空穴去除剂),光照前,将悬浮体在无光条件下搅拌 40 min,达到吸附平衡,然后暴露于氙灯下光照。通过 UV-vis 测定 Cr(Ⅵ)离子溶液的最大吸光度为 365 nm。光催化效率是由以下方程进行计算,光催化效率=C/C_0,其中,C 是 Cr(Ⅵ)浓度在不同时间取样测量的结果,C_0 是初始 Cr(Ⅵ)浓度。所有催化剂测试前均经过 200 ℃活化处理 4 h。

2.3 结果与讨论

合成样品的 XRD 衍射谱如图 2-1 所示,所合成的 TiO$_2$ 前驱体为非晶态结构,基本没有衍射峰。前驱物经过水热反应后获得的介孔 TiO$_2$,所出现的衍射峰均对应锐钛矿 TiO$_2$(JCPDS No.21-1272),其晶格常数为 $a=b=3.785$ Å,$c=9.514$ Å,$\alpha=\beta=\gamma=90°$。衍射峰的展宽表明该样品的晶粒尺寸较小。采用谢乐方程计算结晶尺寸 D:

$$D = 0.9\lambda/(\beta\cos\theta)$$

式中,λ 为 X 射线的波长(1.540 5 Å);β 对应半最大值处的全宽(FWHM);θ 为衍射角。TiO$_2$ 样品的估计尺寸约为 18 nm。合成的 ZIF-8[图 2-1(c)]的衍射峰与模拟的 ZIF-8 和公开的图形[14]相比,具有良好的立方空间群(I43m)[41],表明产物为纯相 ZIF-8。复合材料同时具有纯 ZIF-8 相的衍射峰(2θ=7.4°和 16.6°)和锐钛矿 TiO$_2$ 的衍射峰,表明 ZIF-8 与介孔 TiO$_2$ 形成复合材料。

图 2-2 是制备的 TiO$_2$ 胶体球和介孔 TiO$_2$ 样品的 FE-SEM 图。图 2-2(a)、(b)是 TiO$_2$ 胶体球在不同倍率下的 FE-SEM 图。从图中可以观察到,TiO$_2$ 胶体球呈现均匀球形,表面非常光滑且没有明显的粒状特征,球直径为 (300±50) nm。图 2-2(c)、(d)是 TiO$_2$ 胶体球经过水热处理后获得的介孔

第 2 章　TiO₂/ZIF-8 复合材料的制备及其光催化性能研究

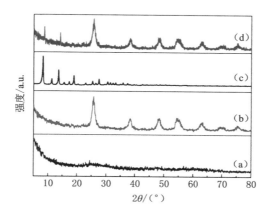

图 2-1　样品的 XRD 图

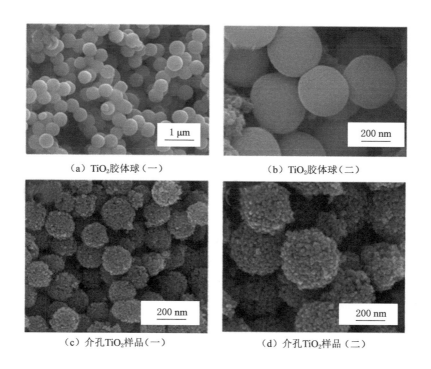

（a）TiO₂胶体球（一）　　　（b）TiO₂胶体球（二）

（c）介孔TiO₂样品（一）　　（d）介孔TiO₂样品（二）

图 2-2　样品的 FE-SEM 图

TiO_2 的 FE-SEM 图。从图 2-2(c)中可以观察到介孔 TiO_2 粒子分散性良好，TiO_2 胶体颗粒水热后表面结晶使微球表面变得粗糙。图 2-2(d)是介孔 TiO_2 的高分辨率 FE-SEM 图，证明该介孔 TiO_2 由尺寸(20 ± 5) nm 均匀纳米晶组成，该结果与 XRD 结果一致。

图 2-3 是制备的纯 ZIF-8 颗粒不同放大倍数的 FE-SEM 图。从图中可以看出，制备的 ZIF-8 为高度均匀的多面体纳米颗粒，纳米颗粒表面光滑，没有其他附着物，且颗粒的尺寸一般为 20～50 nm。这与其他报道制备的 ZIF-8 形貌基本一致。

图 2-3　纯 ZIF-8 颗粒不同放大倍数的 FE-SEM 图

图 2-4 是 ZIF-8/TiO_2 复合材料的 FE-SEM 和 TEM 图。通过原位生长法在 TiO_2 小球表面生长一层 ZIF-8 多面体。从图 2-4(a)中可以看出，ZIF-8 生长后，TiO_2 颗粒尺寸增加，表面变粗糙。图 2-4(b)是高倍率下 ZIF-8/TiO_2 复合材料的 FE-SEM 图。从图中可以看出，在 TiO_2 小球表面有许多多面体颗粒，表明 ZIF-8 粒子已成功地生长在 TiO_2 介孔球表面上，形成了纳米尺寸的 TiO_2/ZIF-8 复合材料。生长在 TiO_2 小球表面上的 ZIF-8 多面体尺寸为 40～70 nm，与单独合成的纯 ZIF-8 晶体相似。图 2-4(c)、(d)是 TiO_2/ZIF-8 复合材料的 TEM 图。由图中可清楚地看出在 TiO_2 小球表面生长粒径约 50 nm 的 ZIF-8 多面体，这进一步证明形成了 TiO_2/ZIF-8 核壳结构。

图 2-5 为 TiO_2、ZIF-8 和 TiO_2/ZIF-8 复合材料的氮气吸附-脱附等温线。根据氮吸附-脱附曲线线性部分(p/p_0=0.1～0.25)计算获得，TiO_2、ZIF-8 和 TiO_2/ZIF-8 的比表面积分别约为 250 m^2/g、1 058 m^2/g 和 397 m^2/g，显然表面经 ZIF-8 粒子修饰的介孔 TiO_2 的比表面积比纯介孔 TiO_2 的更高。ZIF-8 [图 2-5 中的曲线(a)]的曲线是典型的 I 型氮气吸附-脱附等温线[42]，满足

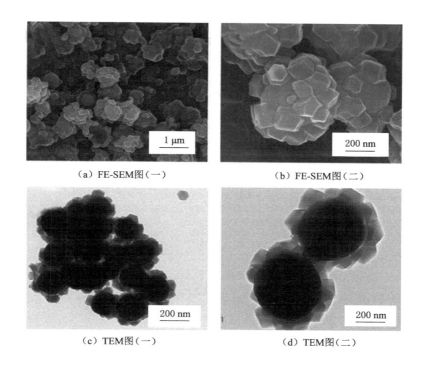

图 2-4 TiO_2/ZIF-8 的 FE-SEM 图和 FEM 图

ZIF-8 微孔骨架的结构特征[37],且在较高的相对压力下(0.8～0.9 MPa),明显的滞后环表明 ZIF-8 晶体的堆积形成大孔隙结构,与以往文献的结果一致[34,37,43]。此外,TiO_2 的氮气吸附-脱附等温线呈现 Ⅳ 型等温线,且具有 H_3 型滞后回线,根据 BDDT 分类,表明 TiO_2 具有介孔特征[44-45]。对比图 2-5 中的(b)、(c)曲线,可知 TiO_2/ZIF-8 的初始吸附量大于 TiO_2 的初始吸附量[42],表明微孔存在于 TiO_2/ZIF-8 纳米球之中,这是因为复合材料中引入了 ZIF-8 粒子。TiO_2/ZIF-8 在高压下的吸附曲线同时具有介孔 TiO_2 和微孔 ZIF-8 颗粒的吸附特征,这表明 TiO_2 与 ZIF-8 形成了复合材料。

为了测定制备的样品的稳定性及 TiO_2/ZIF-8 复合材料中 ZIF-8 的含量,在空气气氛中对 ZIF-8、TiO_2/ZIF-8 和 TiO_2 三个样品进行了 TG-DTA 分析,结果如图 2-6 所示。图 2-6(a)所示为纯 ZIF-8 的 TG 数据。从图中可以看出,当温度达到 600 ℃时,ZIF-8 的总质量损失高达 75%。图 2-6(c)所示为 TiO_2 的 TG 分析结果。当温度达 600 ℃时,TiO_2 的质量损失仅为 1.7%,这主要是

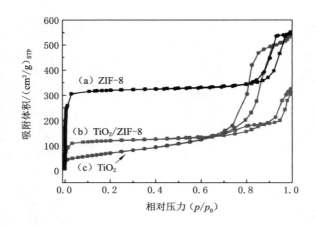

图 2-5　不同样品氮气吸附-脱附等温曲线

由于 TiO_2 表面吸附的水分蒸发所导致的。图 2-6(b)所示为 TiO_2/ZIF-8 的 TG-DTA 数据,可以看出,在整个测试过程中出现三个明显的放热过程,分别是初始阶段中吸附在 TiO_2/ZIF-8 表面的水或甲醇因脱附作用导致质量损失 1.5%;在 250~580 ℃温度范围内,ZIF-8 分子的快速分解而发生明显失重;当温度达到 600 ℃时,ZIF-8 分子已完全转化为 ZnO。在整个测量的温度范围内,TiO_2/ZIF-8 复合材料的总质量损失为 14.18%。根据空气中纯 ZIF-8 的质量损失(75%),可计算出,所制备 TiO_2/ZIF-8 的样品中含有约 20% 的 ZIF-8 和 80% 的 TiO_2。在 TiO_2 上生长的 ZIF-8 实际量小于理论值(理论 ZIF-8 和 TiO_2 的质量比为 1∶2),这是因为,ZIF-8 除了在 TiO_2 表面上生长外,同时也在溶液中结晶沉淀。

图 2-7(a)是 TiO_2、ZIF-8 和 TiO_2/ZIF-8 在 Cr(Ⅵ)水溶液中的光催化还原 Cr(Ⅵ)的性能测试结果。在光照条件下,TiO_2/ZIF-8 纳米球中的 TiO_2 被激发,激发后产生的电子可以转移到 ZIF-8 中,并将 Cr(Ⅵ)还原成 Cr(Ⅲ)。无光和无光催化剂的对照实验中,Cr(Ⅵ)浓度均没有明显降低,这表明光催化还原 Cr(Ⅵ)是在光照和催化剂的共同驱动下进行的。虽然有几种 MOFs 已被证明在光照下具有良好的光催化活性[17,46],但在纯 ZIF-8 作为光催化剂经光照 60 min 后,发现 Cr(Ⅵ)去除率仅为 10%,这是由于 ZIF-8 的弱光催化活性和吸附的协同作用。在光照 60 min 后,纯 TiO_2 对 Cr(Ⅵ)的光催化还原率为 80%,经表面生长 ZIF-8 纳米粒子形成 TiO_2/ZIF-8 纳米球后,相同条件下,Cr(Ⅵ)的去除率高达 99%,其去除效果优于 TiO_2 和 ZIF-8,说明

第 2 章 TiO$_2$/ZIF-8 复合材料的制备及其光催化性能研究

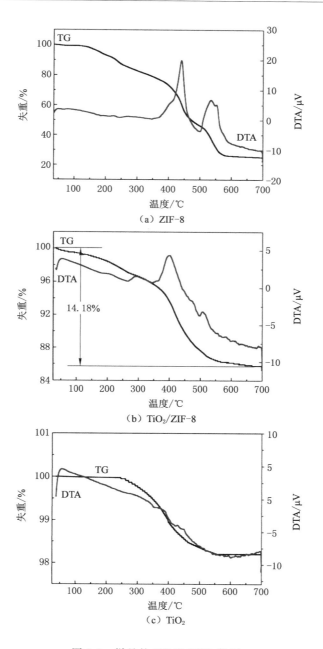

图 2-6 样品的 TG 和 DTA 数据

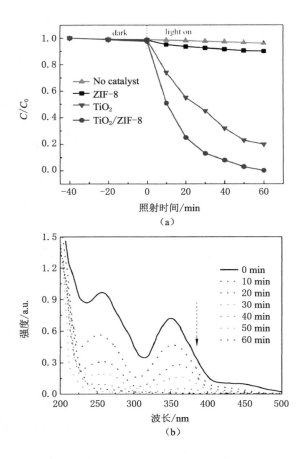

图 2-7 光催化降解 Cr(Ⅵ)的降解曲线和
经不同催化反应时间 Cr(Ⅵ)溶液的吸收光谱图

TiO_2/ZIF-8具有较强的光催化活性,这是由于 ZIF-8 纳米粒子具有高比表面积和多孔结构有利于 Cr(Ⅵ)的吸附和渗透,且在中性条件下(pH=7),ZIF-8 粒子表现出带正电荷[47],而在中性水溶液中 Cr(Ⅵ)的主要形态是 CrO_4^{2-}[34,48]。因此,CrO_4^{2-} 通过静电作用可以有效地吸附于 TiO_2/ZIF-8 光催化剂表面,从而提高 TiO_2/ZIF-8 纳米粒子对 Cr(Ⅵ)的光催化还原性能。图 2-6(b)所示为通过吸收光谱来表征 TiO_2/ZIF-8 光催化还原 Cr(Ⅵ)过程中,Cr(Ⅵ)浓度与时间的变化关系。随着反应时间的增加,Cr(Ⅵ)离子在 365 nm 处的吸收峰强度明显地下降[34,49],在 60 min 时几乎消失,表明 Cr(Ⅵ)

离子被 TiO$_2$/ZIF-8 光催化还原。在紫外光照射 60 min 后的溶液中,Cr(Ⅵ)离子的浓度从初始的 20 mg/L 下降到 0.06 mg/L,溶液变到无色,说明 TiO$_2$/ZIF-8 几乎可以彻底去除溶液中的 Cr(Ⅵ)离子。

值得注意的是,在整个光催化还原过程中,TiO$_2$/ZIF-8 是稳定的。图 2-8 和图 2-9 分别为光催化反应前后 TiO$_2$/ZIF-8 的 XRD 和 SEM 图。从图中可以看出,光催化反应前后的 XRD 衍射峰基本一致,这可说明光催化反应后复合材料的晶体结构并未发生改变。光催化还原 Cr(Ⅵ)后,TiO$_2$/ZIF-8 催化剂的表面形貌也没有发生任何改变,这进一步证明了在催化反应过程中 TiO$_2$/ZIF-8 的稳定性很好。

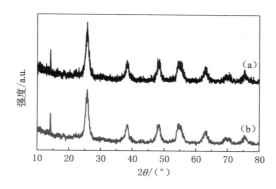

图 2-8　TiO$_2$/ZIF-8 光催化反应前后的 XRD 图

(a) 反应前　　　　　　　　(b) 反应后

图 2-9　TiO$_2$/ZIF-8 光催化反应前后的 SEM 图

光催化剂的稳定性和可重复使用性对光催化的应用非常重要。为了进一步评价 TiO$_2$/ZIF-8 光催化剂的长期性能,采用回收的光催化剂进行辐照光催化还原 Cr(Ⅵ)。每个反应结束后,通过过滤回收使用过的光催化剂,用水和乙醇清洗,真空干燥。TiO$_2$/ZIF-8 光催化还原 Cr(Ⅵ)的循环催化性能如图 2-10(a)所示。

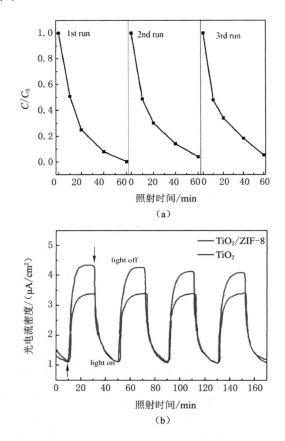

图 2-10　TiO$_2$/ZIF-8 光催化还原 Cr(Ⅵ)的循环催化性能图和
制备的 TiO$_2$/ZIF-8 和 TiO$_2$ 在光照下的瞬态光电流响应图

从图中可以看出,三次循环以后,TiO$_2$/ZIF-8 光催化还原 Cr(Ⅵ)的活性略有下降,主要是由于回收过程中损失了部分催化剂所致。这说明,TiO$_2$/ZIF-8光催化还原 Cr(Ⅵ)的稳定性高。图 2-10(b)是 TiO$_2$ 纳米球和 TiO$_2$/ZIF-8 复合材料在间歇光照条件下的瞬态光电流响应图,TiO$_2$ 纳米球和 TiO$_2$/ZIF-8 的瞬态光电流密度分别为 3.4 μA/cm^2 和 4.3 μA/cm^2,TiO$_2$

表面生长 ZIF-8 后光电流得到显著提高,这表明光致电子空穴对(e^--h^+)的分离效率和光生电荷载流子的寿命都有所提高。对于半导体/MOFs 复合光催化剂,Li 等[32]证明了光生电子可以有效地从半导体转移到 MOFs,这不仅有利于半导体中的电荷分离,而且为吸附在 MOFs 上的分子提供高能电子。因此,TiO_2/ZIF-8 相对于 TiO_2 样品的光催化活性较高,这是由于 ZIF-8 和 TiO_2 纳米球的协同作用,首先,ZIF-8 能有效地吸附水溶液中溶解的 Cr(Ⅵ);其次,TiO_2/ZIF-8 的电荷转移比单独的 TiO_2 更有效。这两方面有效的提升导致 TiO_2/ZIF-8 纳米复合材料具有较高的瞬态光电流密度。

2.4 本章小结

采用原位生长法在 TiO_2 介孔小球表面沉积 ZIF-8 纳米颗粒,制备出 TiO_2/ZIF-8 纳米复合材料。因此,由于 ZIF-8 比 TiO_2 球具有更高的表面积,得到的复合材料具有更高的比表面积。TG-DTA 分析结果表明,所制备的 TiO_2/ZIF-8 复合材料中约含有 20% 的 ZIF-8 和 80% 的 TiO_2。TiO_2/ZIF-8 杂化纳米球对 Cr(Ⅵ)还原的光催化活性比原始 TiO_2 微球更强,这主要是由于复合后光催化剂的比表面积和吸附性能提高;瞬态光电流响应结果表明复合后电子-空穴对的分离效率和载流子的寿命提高。我们预计,在光催化剂中实现 MOF 结构的策略可以应用于其他催化体系,以提高光催化转化效率。

参考文献

[1] RENGARAJ S, VENKATARAJ S, YEON J W, et al. Preparation, characterization and application of Nd-TiO_2 photocatalyst for the reduction of Cr(Ⅵ) under UV light illumination[J]. Applied catalysis B: environmental, 2007, 77(1/2): 157-165.

[2] MOHAN D, PITTMAN C U Jr. Activated carbons and low cost adsorbents for remediation of tri- and hexavalent chromium from water[J]. Journal of hazardous materials, 2006, 137(2): 762-811.

[3] GUPTA V K, RASTOGI A, NAYAK A. Adsorption studies on the removal of hexavalent chromium from aqueous solution using a low cost fertilizer industry waste material[J]. Journal of colloid and interface science, 2010, 342(1): 135-141.

[4] SHI L, WANG T, ZHANG H B, et al. An amine-functionalized iron(Ⅲ) metal-organic framework as efficient visible-light photocatalyst for Cr(Ⅵ) reduction[J]. Advanced science, 2015, 2(3):1500006.

[5] GONG K P, DU F, XIA Z H, et al. Nitrogen-doped carbon nanotube arrays with high electrocatalytic activity for oxygen reduction[J]. Science, 2009, 323(5915):760-764.

[6] MOHAPATRA P, SAMANTARAY S K, PARIDA K. Photocatalytic reduction of hexavalent chromium in aqueous solution over sulphate modified titania[J]. Journal of photochemistry and photobiology A: chemistry, 2005, 170(2):189-194.

[7] WANG L, KANG S Z, LI X Q, et al. Rapid and efficient photocatalytic reduction of hexavalent chromium by using "water dispersible" TiO_2 nanoparticles[J]. Materials chemistry and physics, 2016, 178:190-195.

[8] LIU X J, PAN L K, LV T, et al. Microwave-assisted synthesis of CdS-reduced graphene oxide composites for photocatalytic reduction of Cr(Ⅵ)[J]. Chemical communications (Cambridge, England), 2011, 47(43): 11984-11986.

[9] ZHANG Y C, LI J, ZHANG M, et al. Size-tunable hydrothermal synthesis of SnS_2 nanocrystals with high performance in visible light-driven photocatalytic reduction of aqueous Cr(Ⅵ)[J]. Environmental science and technology, 2011, 45(21):9324-9331.

[10] YANG W L, ZHANG L, HU Y, et al. Microwave-assisted synthesis of porous Ag_2S-Ag hybrid nanotubes with high visible-light photocatalytic activity[J]. Angewandte chemie international edition, 2012, 51(46): 11501-11504.

[11] TONG H, OUYANG S X, BI Y P, et al. Nano-photocatalytic materials: possibilities and challenges[J]. Advanced materials, 2012, 24(2): 229-251.

[12] CHANG X X, WANG T, GONG J L. CO_2 photo-reduction: insights into CO_2 activation and reaction on surfaces of photocatalysts[J]. Energy and environmental science, 2016, 9(7):2177-2196.

[13] TU W G, ZHOU Y, ZOU Z G. Photocatalytic conversion of CO_2 into renewable hydrocarbon fuels: state-of-the-art accomplishment, challenges, and

prospects[J].Advanced materials,2014,26(27):4607-4626.

[14] PARK K S,NI Z,CÔTÉ A P,et al.Exceptional chemical and thermal stability of zeolitic imidazolate frameworks[J].Proceedings of the national academy of sciences of the United States of America,2006,103(27):10186-10191.

[15] HUANG G,CHEN Y Z,JIANG H L.Metal-organic frameworks for catalysis[J].Acta chimica sinica,2016,74(2):113.

[16] CAVKA J H,JAKOBSEN S,OLSBYE U,et al.A new zirconium inorganic building brick forming metal organic frameworks with exceptional stability[J].Journal of the American chemical society,2008,130(42):13850-13851.

[17] FU Y H,SUN D R,CHEN Y J,et al.An amine-functionalized titanium metal-organic framework photocatalyst with visible-light-induced activity for CO_2 reduction[J].Angewandte chemie,2012,124(14):3420-3423.

[18] XU H Q,HU J H,WANG D K,et al.Visible-light photoreduction of CO_2 in a metal-organic framework: boosting electron-hole separation via electron trap states[J].Journal of the American chemical society,2015,137(42):13440-13443.

[19] DHAKSHINAMOORTHY A,ASIRI A M,GARCÍA H.Metal-organic framework(MOF) compounds: photocatalysts for redox reactions and solar fuel production[J].Angewandte chemie international edition,2016,55(18):5414-5445.

[20] LAURIER K G M,VERMOORTELE F,AMELOOT R,et al.Iron(III)-based metal-organic frameworks as visible light photocatalysts[J].Journal of the American chemical society,2013,135(39):14488-14491.

[21] LI Y,XU H,OUYANG S X,et al.Metal-organic frameworks for photocatalysis[J].Physical chemistry chemical physics,2016,18(11):7563-7572.

[22] WANG D K,WANG M T,LI Z H.Fe-based metal-organic frameworks for highly selective photocatalytic benzene hydroxylation to phenol[J].ACS catalysis,2015,5(11):6852-6857.

[23] GOH T W,XIAO C X,MALIGAL-GANESH R V,et al.Utilizing mixed-linker zirconium based metal-organic frameworks to enhance the visible light photocatalytic oxidation of alcohol[J].Chemical engineering

science, 2015, 124: 45-51.

[24] SHI L, WANG T, ZHANG H B, et al. An amine-functionalized iron(Ⅲ) metal-organic framework as efficient visible-light photocatalyst for Cr(Ⅵ) reduction[J]. Advanced science, 2015, 2(3): 1500006.

[25] LIANG R W, SHEN L J, JING F F, et al. NH_2-mediated indium metal-organic framework as a novel visible-light-driven photocatalyst for reduction of the aqueous Cr(Ⅵ)[J]. Applied catalysis B: environmental, 2015, 162: 245-251.

[26] LIANG R W, JING F F, SHEN L J, et al. MIL-53(Fe) as a highly efficient bifunctional photocatalyst for the simultaneous reduction of Cr(Ⅵ) and oxidation of dyes[J]. Journal of hazardous materials, 2015, 287: 364-372.

[27] MAHATA P, MADRAS G, NATARAJAN S. Novel photocatalysts for the decomposition of organic dyes based on metal-organic framework compounds[J]. The journal of physical chemistry B, 2006, 110(28): 13759-13768.

[28] WANG S B, LIN J L, WANG X C. Semiconductor-redox catalysis promoted by metal-organic frameworks for CO_2 reduction[J]. Physical chemistry chemical physics: PCCP, 2014, 16(28): 14656-14660.

[29] SU Y, ZHANG Z, LIU H, et al. $Cd_{0.2}Zn_{0.8}S$@UiO-66-NH_2 nanocomposites as efficient and stable visible-light-driven photocatalyst for H_2 evolution and CO_2 reduction[J]. Applied catalysis B: environmental, 2017, 200: 448-457.

[30] CHANG N, HE D Y, LI Y X, et al. Fabrication of TiO_2@MIL-53 core-shell composite for exceptionally enhanced adsorption and degradation of nonionic organics[J]. RSC advances, 2016, 6(75): 71481-71484.

[31] ZENG X, HUANG L Q, WANG C N, et al. Sonocrystallization of ZIF-8 on electrostatic spinning TiO_2 nanofibers surface with enhanced photocatalysis property through synergistic effect[J]. ACS applied materials and interfaces, 2016, 8(31): 20274-20282.

[32] LI R, HU J H, DENG M S, et al. Integration of an inorganic semiconductor with a metal-organic framework: a platform for enhanced gaseous photocatalytic reactions[J]. Advanced materials, 2014, 26(28): 4783-4788.

[33] ZHAN W W, KUANG Q, ZHOU J Z, et al. Semiconductor@metal-organic framework core-shell heterostructures: a case of ZnO@ZIF-8 nanorods with selective photoelectrochemical response[J]. Journal of the American chemical society, 2013, 135(5): 1926-1933.

[34] WANG X B, LIU J, LEONG S, et al. Rapid construction of ZnO@ZIF-8 heterostructures with size-selective photocatalysis properties[J]. ACS applied materials and interfaces, 2016, 8(14): 9080-9087.

[35] YUE X X, GUO W L, LI X H, et al. Core-shell Fe_3O_4@MIL-101(Fe) composites as heterogeneous catalysts of persulfate activation for the removal of Acid Orange 7[J]. Environmental science and pollution research international, 2016, 23(15): 15218-15226.

[36] WANG R, GU L N, ZHOU J J, et al. Quasi-polymeric metal-organic framework UiO-66/g-C_3N_4 Heterojunctions for enhanced photocatalytic hydrogen evolution under visible light irradiation [J]. Advanced materials interfaces, 2015, 2(10): 1500037.

[37] LIU Q, LOW Z X, LI L X, et al. ZIF-8/Zn_2GeO_4 nanorods with an enhanced CO_2 adsorption property in an aqueous medium for photocatalytic synthesis of liquid fuel[J]. Journal of materials chemistry A, 2013, 1(38): 11563.

[38] WANG M T, WANG D K, LI Z H. Self-assembly of CPO-27-Mg/TiO_2 nanocomposite with enhanced performance for photocatalytic CO_2 reduction[J]. Applied catalysis B: environmental, 2016, 183: 47-52.

[39] HUANG X C, LIN Y Y, ZHANG J P, et al. Ligand-directed strategy for zeolite-type metal-organic frameworks: zinc(Ⅱ) imidazolates with unusual zeolitic topologies[J]. Angewandte chemie international edition, 2006, 45(10): 1557-1559.

[40] JIANG X, HERRICKS T, XIA Y. Monodispersed spherical colloids of titania: synthesis, characterization, and crystallization[J]. Advanced materials, 2003, 15(14): 1205-1209.

[41] TRAN U P N, LE K K A, PHAN N T S. Expanding applications of Metal-Organic frameworks: zeolite imidazolate framework ZIF-8 as an efficient heterogeneous catalyst for the Knoevenagel reaction[J]. ACS catalysis, 2011, 1(2): 120-127.

[42] KRUK M,JARONIEC M.Gas adsorption characterization of ordered organic-inorganic nanocomposite materials[J].Chemistry of materials,2001,13(10):3169-3183.

[43] HE M,YAO J F,LIU Q,et al.Facile synthesis of zeolitic imidazolate framework-8 from a concentrated aqueous solution[J].Microporous and mesoporous materials,2014,184:55-60.

[44] YAN S C,OUYANG S X,GAO J,et al.A room-temperature reactive-template route to mesoporous $ZnGa_2O_4$ with improved photocatalytic activity in reduction of CO_2[J].Angewandte chemie,2010,122(36):6544-6548.

[45] LOU X W,DENG D,LEE J Y,et al.Thermal formation of mesoporous single-crystal Co_3O_4 nano-needles and their lithium storage properties[J].Journal of materials chemistry,2008,18(37):4397.

[46] WANG C,XIE Z G,DEKRAFFT K E,et al.Doping metal-organic frameworks for water oxidation,carbon dioxide reduction,and organic photocatalysis[J].Journal of the American chemical society,2011,133(34):13445-13454.

[47] KHAN N A,JUNG B K,HASAN Z,et al.Adsorption and removal of phthalic acid and diethyl phthalate from water with zeolitic imidazolate and metal-organic frameworks[J].Journal of hazardous materials,2015,282:194-200.

[48] WANG X B,CAI W P,LIN Y X,et al.Mass production of micro/nano-structured porous ZnO plates and their strong structurally enhanced and selective adsorption performance for environmental remediation[J].Journal of materials chemistry,2010,20(39):8582.

[49] YU J Y,ZHUANG S D,XU X Y,et al.Photogenerated electron reservoir in hetero-p-n CuO-ZnO nanocomposite device for visible-light-driven photocatalytic reduction of aqueous Cr(Ⅵ)[J].Journal of materials chemistry A,2015,3(3):1199-1207.

第3章　ZIF-8/Zn_2GeO_4 纳米棒的制备及其光催化还原 CO_2 研究

3.1　引言

由于化石燃料的持续燃烧，大气中的 CO_2 浓度持续上升，已成为严重的全球环境问题。为了开发可靠、经济有效且可以减少 CO_2 排放的技术，已经进行了大量的研究工作。而 CO_2 捕获与排放储存(CCS)正在被广泛地研究，利用太阳能光催化将 CO_2 转化为碳氢燃料，为利用 CO_2 进行可再生碳固定和能源储存提供了一个强有力的解决方案。许多金属氧化物半导体如 ZnO[1]、TiO_2[2-4]、CdS[1,5]、$SrTiO_3$[6]、$W_{18}O_{49}$[7]、Zn_2SnO_4[8]、$ZnGa_2O_4$[9] 和 Zn_2GeO_4[10-12] 已被研究用于 CO_2 光还原的燃料，但它们的 CO_2 转化效率很低，根本原因是：① 传统的半导体光催化剂对 CO_2 吸附性能差，CO_2 分子难以活化；② 光生载流子复合严重，寿命短[13]。因此，如何实现 CO_2 在光催化剂表面的富集和活化、抑制光生载流子的复合、促进光生电子和空穴的有效分离成为提高光催化还原 CO_2 转换效率的重要挑战[14]。增加光催化剂的比表面积已被证明是提供更多的反应位点和更好的吸附性能从而提高光催化活性的有效途径。沸石和介孔材料具有极高的比表面积、规则有序的孔道结构、狭窄的孔径分布、孔径大小连续可调等特点，使得其在光催化还原 CO_2 反应中备受关注。如钛沸石、Ti-MCM-41、Ti-MCM-48、ZrCu（Ⅰ）-MCM-41 和 Ti-SBA-15[15-19]，被证明可以提高产物对 CO_2 的吸附和选择性。但是，进一步提高这类光催化活性是非常有限的，因为这些无机半导体光催化剂需要综合考虑高的结晶度和比表面积。

金属有机骨架材料(MOFs)由于具有高比表面积、大孔径、良好且界限明确的纳米级孔腔和化学定制能力等优点，作为吸附剂和催化剂的潜力已经引起了越来越多学者的研究兴趣[20]。MOFs 被认为是很有前途的 CO_2 捕获和储存(CCS)的候选材料[21-25]。一些 MOFs 也被证明具有光催化活性。例如,一种氨功能化钛[NH_2-MIL-125(Ti)]光催化剂被合成用于可见光下的 CO_2 还原[26],但其 CO_2 还原效率低于传统的半导体光催化剂体系。此外,一些 MOFs,如沸石咪唑框架(ZIFs),除了比沸石和介孔材料具有更高的 CO_2 吸附性能外,还具有优异的热化学稳定性和在水中的结构稳定性。为此,在半导体光催化剂中引入 MOFs 可以通过增强光催化反应中 CO_2 和中间体的吸附来改善 CO_2 的光还原效率;所得到的杂化催化剂结合了 MOFs 的高吸附性能和无机半导体的高稳定性、能带可调性、低成本和环境友好性的优点。截至 2012 年,还没有关于这种半导体-MOF 纳米复合材料光催化还原 CO_2 的报道。在此,我们报道了通过在半导体纳米棒上生长 MOF 纳米颗粒来合成半导体-MOF 纳米复合材料的方法,该方法可以显著增强水介质中 CO_2 的光催化转化为液态燃料(CH_3OH)的能力。本研究选择 ZIF-8 和 Zn_2GeO_4 制备半导体-MOF 纳米复合光催化剂。Zn_2GeO_4 是一种重要的半导体光催化剂,在降解有机污染物方面得到了广泛的研究[27]。ZIF-8 具有直径为 11.6 Å 的笼状孔和直径为 3.4 Å 的窗口孔,在水中稳定性好。如图 3-1 所示,采用简单的水热法在氢氧化四甲基铵(TMAOH)水溶液中合成了 Zn_2GeO_4 纳米棒;然后在这些纳米棒上生长 ZIF-8 纳米颗粒,得到 Zn_2GeO_4/ZIF-8 复合纳米棒[28-29]。虽然 MOFs 和 ZIFs 的 CO_2 气体吸附-解吸特性已经得到了很好的研究[30],但迄今为止,对 MOFs 和 ZIFs 在水介质中的溶解气体吸附行为的研究很少。本研究结果表明,ZIF-8 对水中溶解的 CO_2 具有良好的吸附性能。独特的 Zn_2GeO_4/ZIF-8 纳米复合材料既具有 ZIF-8 纳米颗粒的大比表面积、高 CO_2 吸附能力,又具有 Zn_2GeO_4 纳米棒的高结晶度和光催化活性,使得 CO_2 光催化转化为 CH_3OH 的能力比纯 Zn_2GeO_4 纳米棒提高了 62%。

3.2 实验部分

3.2.1 Zn_2GeO_4 纳米棒的制备

选用的所有的化学制品均为分析级,不需进一步纯化即可使用。典型的 Zn_2GeO_4 纳米棒制备方案如下:将 0.52 g GeO_2(2.5 mmol)和 1.10 g $Zn(CH_3COO)_2 \cdot 2H_2O$(5 mmol)加入 25% 四甲基氢氧化亚胺(TMAOH)水溶液(15 mL)中。将混合溶液搅拌 40 min,然后转移到内容量为 25 mL 聚四

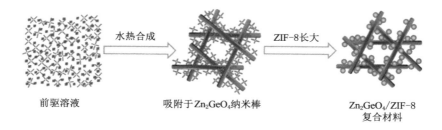

图 3-1 Zn_2GeO_4/ZIF-8 合成示意图

氟乙烯内衬的不锈钢高压釜中,在 180 ℃ 烘箱中水热反应 12 h,然后自然冷却到室温。打开反应釜,将沉淀物离心分离,依次用去离子水、无水乙醇各洗 3 次,60 ℃ 干燥 12 h,得到白色的 Zn_2GeO_4 粉末。

3.2.2 ZIF-8 纳米颗粒的合成

ZIF-8 纳米颗粒是按照先前报道的方法制备的[31]。分别取 0.587 g $Zn(NO_3)_2 \cdot 6H_2O$ 和 1.298 g 2-甲基咪唑(Hmin)溶于 40 mL 甲醇(MeOH)中,磁搅拌下将 Hmim 溶液倒入 $Zn(NO_3)_2 \cdot 6H_2O$ 溶液中。然后室温下搅拌混合溶液 2 h,离心将固体产物从乳状胶体分散体中分离出来。离心得到的白色 ZIF-8 产品用无水乙醇洗涤,离心 3 次,60 ℃ 干燥备用。

3.2.3 Zn_2GeO_4/ZIF-8 的制备

将 0.4 g 制备的 Zn_2GeO_4 纳米棒和 0.587 g $Zn(NO_3)_2 \cdot 6H_2O$ 加入 40 mL MeOH 中搅拌 1 h,得到溶液 1;将 1.298 g Hmim 溶解于 40 mL MeOH 中(Zn_2GeO_4 与 ZIF-8 的理论质量比为 1∶2)得到溶液 2。将溶液 2 倒入溶液 1 中,磁力搅拌反应 1 h,然后静置约 5 min,将上部悬浮液层倒掉,得到的白色沉淀用水和酒精清洗几次,然后放入烘箱中 60 ℃ 干燥 12 h。

3.2.4 样品表征

用日本理学公司生产的 Ultima Ⅲ 型 X 射线衍射仪对产物进行物相分析,以 Cu-Kα 为辐射源,波长为 1.541 78 Å,采用连续扫描方式,扫描速度为 5°/min,扫描范围 $2\theta = 5° \sim 80°$。比表面积大小通过美国 Micromeritics 公司生产的 TriStar 3000 型比表面-孔径分布分析仪器测定,N_2 的吸附/脱附是在 77 K 温度下完成的,用 BET 方法计算出其比表面积。扫描电镜(SEM)照片采用荷兰 FEI 公司生产的 Tecnai G2 F30 S-Twin 型场发射扫描电镜（FE-SEM）拍摄，加速电压设置为 15 kV。透射电镜（TEM）照片和高分辨透射电镜（HR-TEM）照片用日本 JEOL 公司生产的 JEM-3010 型透射电镜获得，工作电压为 200 kV。样品的成分分析采用赛默飞世尔科技 K-Alpha 型 X 射线光电子能谱（XPS）测试，以 C 1s 结合能 284.8 eV 为参比校正各元素的电子结合能。在 Netzsch STA 449 F1 热分析仪上同时进行了热重分析（TG）和差热分析（DTA）。为此，6 mg 的样品装到氧化铝坩埚中，空气中以 5 ℃/min 的速度从室温加热到 700 ℃。CO_2 吸附-脱附等温线测定在 273 K 下使用自动体积吸附装置（Micromertics ASAP 2010）。紫外-可见反射光谱由日本岛津公司生产的 UV-2500PC 型紫外-可见分光光度计分析获得，通过 Kubelka-Munk 方法转换成吸收光谱。用岛津 TOC-LCSH/CSN&amp 测定水溶液中总溶解的碳浓度。

3.2.5 CO_2 光催化还原表征

Pt 助催化剂的担载采用光催化还原法，$H_2PtCl_6 \cdot 6H_2O$ 为 Pt 源[10,32]。典型的实验条件为：将 0.2 g 光催化剂加入一个玻璃容器中，加入 40 mL 蒸馏水、15 mL 的甲醇和一定量的氯铂酸溶液。将配置好的溶液磁力搅拌形成悬浮溶液，用 300 W 氙灯光照 8 h，使得 Pt 还原并在光催化剂表面沉积。反应结束后，过滤反应溶液，并用水和酒精多次洗涤，然后放入 60 ℃ 的鼓风干燥箱中干燥 12 h。光催化反应前，催化剂在 200 ℃ 真空下脱气 12 h，以去除材料表面的吸附物或残留溶剂。光催化还原 CO_2 反应中，将 0.20 g 光催化剂均匀分散在 100 mL 亚硫酸钠（Na_2SO_3，0.10 mol/L）溶液中。将高纯 N_2 脱气通入溶液中除去溶液中的氧，然后将高纯 CO_2 气体鼓入溶液中，搅拌 90 min 至 CO_2 吸附-脱附平衡。以 500 W 的氙灯作为光源，光催化反应期间，每间隔一段时间从反应溶液中取约 1 mL 的溶液，离心后注入气相色谱仪（GC）进行 CH_3OH 浓度分析，GC 采用 FID 检测器和 Agilent G1888 顶空采样器（Agilent 7890A）。CH_3OH 的生成定义为 CH_3OH 的总产量除以反应时间。

3.3 结果与讨论

样品的 X 射线衍射（XRD）结果如图 3-2 所示。在纯 Zn_2GeO_4 纳米棒的 XRD 图谱[图 3-2(a)]中，所有衍射峰与四方相的 Zn_2GeO_4（JCPDS 11-0687）的衍射峰一一对应，晶格常数为 $a=b=1.423$ nm，$c=0.953$ nm，$\alpha=\beta=90°$，

$\gamma=120°$。图中没有观察到其他溶液制备方法中容易出现的 ZnO 和 GeO_2 杂相衍射峰的存在,说明该实验条件能够合成完全纯相的 Zn_2GeO_4 样品。合成的纯 ZIF-8 的 XRD 图谱[图 3-2(b)]与报道的图谱吻合[20],这表明产物主要为结晶的 ZIF-8。在 Zn_2GeO_4 上沉积 ZIF-8 颗粒后[图 3-2(c)],在 $2\theta=7.4°$、$16.6°$、$18.0°$ 和 $26.6°$ 处出现了新的峰值,这与 ZIF-8 的衍射峰刚好一一对应。没有发现其他物相的特征峰,说明制备的样品仅由 Zn_2GeO_4 和 ZIF-8 组成。

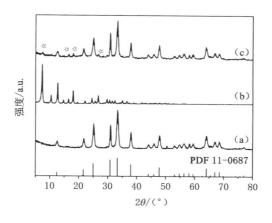

图 3-2　样品的 XRD 图

图 3-3 为该纯 Zn_2GeO_4 样品的 SEM 图。从图 3-3(a)中可以看出,制备的样品颗粒大小均匀,形状规整,具有单分散的棒状结构。纳米棒表面光滑,单个纳米棒的长度为 200~300 nm,均匀直径为 20~50 nm。

图 3-3　不同放大倍数下 Zn_2GeO_4 纳米棒的 SEM 图

对采用室温沉淀法制备的 ZIF-8 采用 SEM 对其进行形貌分析,如图 3-4 所示。从图中可以看出,制备的 ZIF-8 为高度均匀的纳米颗粒,颗粒为典型的菱形十二面体形状,有明显的棱角,颗粒表面光滑,没有明显附着物和缺陷。通过标尺对比可知 ZIF-8 粒径为 50～100 nm。采用 TEM 进一步对制备的 ZIF-8 晶种进行表征,结果如图 3-4(b)插图所示,从图中可以看出,纳米颗粒的 TEM 投影为多边形形状,这说明制备的 ZIF-8 为多面体形状,颗粒尺寸为 50～100 nm,与 SEM 结果一致。颗粒表面没有任何附着物。

图 3-4　不同放大倍数下 ZIF-8 的 FE-SEM 图

在 Zn_2GeO_4 表面沉积 ZIF-8 后,对制备的复合材料进行 SEM 和 TEM 分析,结果如图 3-5 所示。由图 3-5(a)可知,ZIF-8 沉积后,光滑的 Zn_2GeO_4 纳米棒出现很多细小的颗粒,形成链状结构。高放大率 FE-SEM 图像表明 Zn_2GeO_4 纳米棒表面有许多规则的纳米颗粒,纳米颗粒的形貌与纯 ZIF-8 类似,这证明 Zn_2GeO_4 纳米棒上成功生长了 ZIF-8 纳米颗粒。纳米棒上的 ZIF-8 纳米颗粒尺寸为 10～50 nm,比纯 ZIF-8 纳米颗粒要小(图 3-4)。

实验中发现,控制沉积时间可以很好地控制 Zn_2GeO_4/ZIF-8 中 ZIF-8 的含量。在合成初期(即 10 min),如图 3-6 所示,Zn_2GeO_4 纳米棒上生长的纳米颗粒很少。随着反应时间的增加,Zn_2GeO_4 纳米棒上形成了更多、更大的 ZIF-8 纳米颗粒[图 3-6(b)、(c)、(d)]。Zn_2GeO_4/ZIF-8 纳米线的 TEM 图如图 3-5(c)、(d)所示。虽然 ZIF-8 对电子束非常敏感,但可以清楚地看到修饰在 Zn_2GeO_4 纳米棒表面约 10 nm 离散分布的 ZIF-8 纳米颗粒[图 3-5(c)]。HRTEM 图像显示,纳米棒的晶格条纹清晰,晶面间距 d 约为 4.10 Å,对应于四方相 Zn_2GeO_4 的(300)晶面[图 3-5(d)]。所选区域电子衍射(SAED)图表明,Zn_2GeO_4 纳米棒长度为[001]方向、宽度为[100]方向,这表明纳米棒的生

第3章 ZIF-8/Zn_2GeO_4 纳米棒的制备及其光催化还原 CO_2 研究

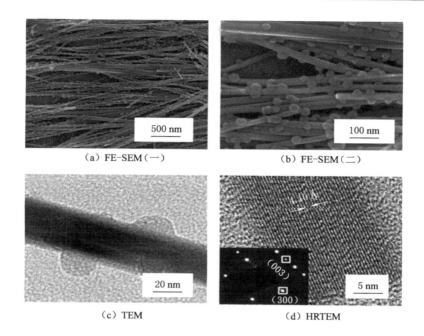

图 3-5　Zn_2GeO_4/ZIF-8 的 FE-SEM 图、TEM 图和 HRTEM 图

长主要是沿着 c 轴方向进行的。

图 3-7 所示为制备的 Zn_2GeO_4、ZIF-8 和 Zn_2GeO_4/ZIF-8 的 X 射线光电子能谱(XPS)图谱,除了 Zn、O、Ge、N 和 C 元素的峰外,未出现其他元素的峰。Zn_2GeO_4 纳米棒的 402.68 eV 处的 N 1s 峰归属于 TMAOH 的烷基铵中的 N[33],这证实了在制备的 Zn_2GeO_4 纳米棒表面吸附了 TMAOH。N 1s 在纯 ZIF-8 和 Zn_2GeO_4/ZIF-8 中的位置分别为 398.58 eV 和 398.68 eV,这分别归于 C—N 键和 2-甲基咪唑中的 N[34],这进一步验证了 Zn_2GeO_4/ZIF-8 中 ZIF-8 的存在。Zn、Ge、O、N 峰强度的显著差异说明了三个样品中元素含量的不同。

样品的红外光谱(FT-IR)(图 3-8)表明,纯 Zn_2GeO_4 纳米棒在 3 300~3 500 cm^{-1} 之间出现一个吸收峰,对应着—NH_2 的伸缩振动,进一步说明了 Zn_2GeO_4 纳米棒表面存在—NH_2。在 Zn_2GeO_4 纳米棒上原位沉积 ZIF-8 后,出现了一些新的 FT-IR 峰。在 1 146 cm^{-1} 和 1 308 cm^{-1} 处的吸收带被认为是碳氢化合物的振动,而在 1 384 cm^{-1} 处的峰值被认为是碳氢化合物的伸缩。因此,FT-IR 光谱也证实了 ZIF-8/Zn_2GeO_4 纳米棒的杂化结构。值得注意的是,功能化烷基铵对 Zn_2GeO_4 纳米棒上的 ZIF-8 纳米粒子的生长起着至关重

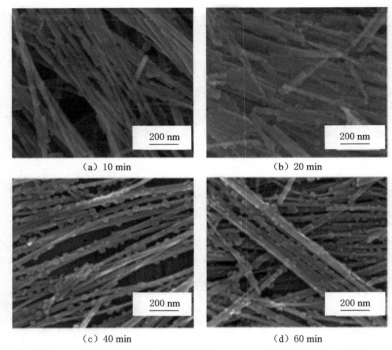

图 3-6 不同 ZIF-8 沉积时间制备的 Zn_2GeO_4/ZIF-8 FE-SEM 图

要的作用,而在氟离子合成溶液中,Zn_2GeO_4 纳米棒上并没有生长 ZIF-8 纳米粒子(图 3-9)。

采用热重(TG)评价了制备的样品的热稳定性。对于纯 Zn_2GeO_4 纳米棒,TG 分析结果表明,其不存在明显的分解过程,加热到 700 ℃时质量损失仅为 0.5%[图 3-10(a)],这个失重是吸附在纳米棒表面的 TMAOH 和 H_2O 的解吸造成的。ZIF-8 的 TG 结果表明,其在 200 ℃ 以下失重约为 11.8%,这是由于 ZIF-8 孔腔中溶剂分子 MeOH 和纳米晶体表面吸附的反应配体 2-甲基咪唑的脱附所引起的。ZIF-8 在 200～500 ℃ 的温度范围失重明显,700 ℃ 时总失重约为 75%。合成的 Zn_2GeO_4/ZIF-8 纳米晶的 TG/DTA 曲线与 ZIF-8 纳米晶相似。Zn_2GeO_4/ZIF-8 试样在测温范围内的总失重约为 18%,可以计算出杂交试样中 ZIF-8 的含量约为 25%,Zn_2GeO_4 的含量约为 75%。Zn_2GeO_4 上实际生长的 ZIF-8 的量小于理论值(理论值 ZIF-8/Zn_2GeO_4 质量比为 1∶2)。这可以解释为除了 Zn_2GeO_4 上的 ZIF-8 生长外,ZIF-8 还在溶液中结晶沉淀,该过程消耗了溶液中的反应物。

第 3 章 ZIF-8/Zn₂GeO₄ 纳米棒的制备及其光催化还原 CO₂ 研究

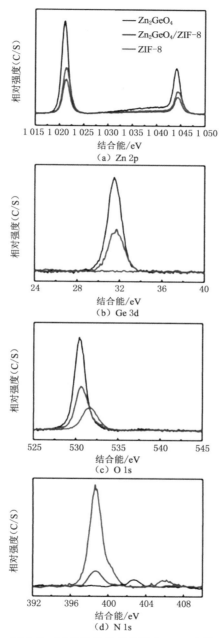

图 3-7 制备样品的高分辨率 XPS 光谱

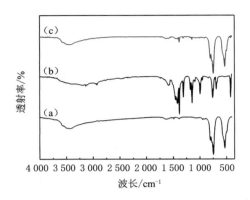

图 3-8 FT-IR 光谱图

(a) FE-SEM(一) (b) FE-SEM(二)

(c) TEM (d) HRTEM

图 3-9 氟离子体系中 Zn_2GeO_4 纳米棒的 SEM 图及
ZIF-8 原位生长后 Zn_2GeO_4 的 SEM 图

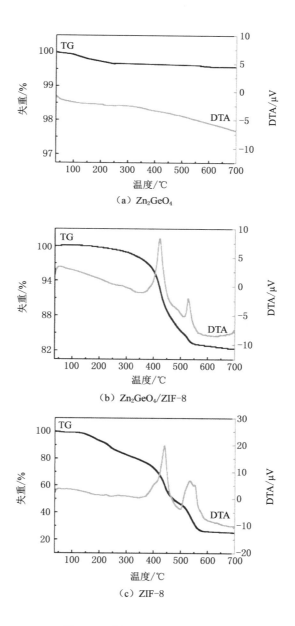

图 3-10 样品的 TG-DTA 曲线

采用 N_2 吸附-脱附测试对 Zn_2GeO_4、ZIF-8 和 Zn_2GeO_4/ZIF-8 样品进行了分析。N_2 吸附-脱附结果表明[图 3-11(a)]，Zn_2GeO_4/ZIF-8 杂交纳米棒中由于微孔的存在，在很低的相对压力下，Zn_2GeO_4/ZIF-8 对 N_2 的吸附量明显增加。由 N_2 吸附-脱附计算得到 Zn_2GeO_4、ZIF-8 和 Zn_2GeO_4/ZIF-8 的比表面积分别为 40.9 m^2/g、1 160.1 m^2/g 和 319.5 m^2/g。Zn_2GeO_4/ZIF-8 样品的高比表面积归功于纳米棒上生长的 ZIF-8 颗粒。通过比较三个样品的 CO_2 吸附等温曲线，可以看出 Zn_2GeO_4/ZIF-8 纳米棒对 CO_2 的吸附能力高于 Zn_2GeO_4 纳米棒，原因是 ZIF-8 纳米颗粒对 CO_2 的吸附能力较强[图 3-11(b)]。在 1 atm 和 273 K 时，ZIF-8、Zn_2GeO_4/ZIF-8 和 Zn_2GeO_4 的最大 CO_2 吸收量分别为 34.5 cm^3/g、15.5 cm^3/g 和 4.9 m^2/g。我们还测定了室温下 ZIF-8、Zn_2GeO_4/ZIF-8 和 Zn_2GeO_4 在低 CO_2 浓度水溶液中的吸附能力。将高纯度 CO_2 气体经去离子水浸泡 1 h，制得无机碳溶解浓度为 184.2 mg/L 的 CO_2 水溶液。将 0.5 g 的样品 ZIF-8、Zn_2GeO_4/ZIF-8 或 Zn_2GeO_4 加入 50 mL CO_2 水溶液中，在 500 r/mim 下搅拌 1.5 h，CO_2 浓度分别降至 20.2 mg/L、60.3 mg/L 和 123.4 mg/L。在对照实验中，将 50 mL 不含任何粉末的 CO_2 溶液在 500 r/min 下搅拌 1.5 h，发现溶解的无机碳浓度下降到 146.0 mg/L。计算得出 ZIF-8、Zn_2GeO_4/ZIF-8 和 Zn_2GeO_4 对溶解 CO_2 的吸附能力分别为 251.6 mg/g、171.4 mg/g 和 45.2 mg/g。Zn_2GeO_4/ZIF-8 对溶解 CO_2 的吸附能力是 Zn_2GeO_4 的 3.8 倍。

一般来说，以水蒸气为还原剂，利用半导体光催化剂在复杂的气-固体系中进行 CO_2 光催化还原的主要产物是 CH_4[35]。本实验中，采用 CO_2 水溶液为反应物，将 Zn_2GeO_4 和 Zn_2GeO_4/ZIF-8 加入 CO_2 水溶液中进行光催化还原，以获得更高价值的化学物质，如甲醇而不是 CH_4。据我们所知，这也是第一次研究 CO_2 与 Zn_2GeO_4 在水溶液中的光还原性能。正如预期的那样，主要还原产物被确定为甲醇。甲烷、乙醇和其他碳氢化合物可能会少量形成，因为含量太低，所以检测不到。对于 Zn_2GeO_4 纳米棒光催化剂，反应初始时，随着反应时间的增加，CH_3OH 的产量增加；但是在反应时间为 6～10 h 时，CH_3OH 的产量趋于稳定，基本没有变化，这可能是由于光催化反应过程中生成的中间产物在光催化剂表面有较强的吸附作用，覆盖了催化剂的活性位点，使得催化反应难以进行。这种情况可以将使用的催化剂在真空下通过热处理重新活化。由图 3-12 可知，在光照 10 h 内，Zn_2GeO_4 纳米棒的 CH_3OH 产率约为 1.43 μmol/g[曲线(a)]。在没有光催化剂的情况下进行的 CO_2 还原实验中没有出现甲醇，证明了在光催化剂的作用下，CO_2 还原

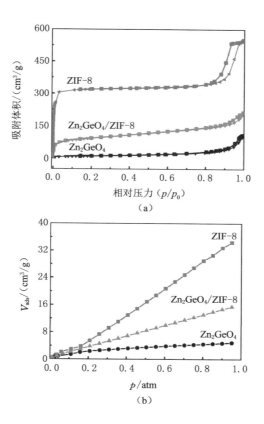

图 3-11 制备的样品的 N_2 吸附-脱附曲线及 CO_2 吸附曲线（273 K）

反应是由光驱动的。虽然最近有一些 MOFs 在光照下显示出了较低的 CO_2 光催化活性[36-38]，但是使用纯 ZIF-8 作为光催化剂光催化还原 CO_2 没有检测到有机产物，这说明 ZIF-8 本身没有光催化还原 CO_2 活性。用 ZIF-8 纳米颗粒修饰 Zn_2GeO_4 纳米棒后，Zn_2GeO_4/ZIF-8 光催化剂 CH_3OH 的产率大大提高，如图 3-12(b)所示。CH_3OH 的总产量随着光照时间的增加而增加，实验中连续照射 11 h 得到的 CH_3OH 总产量为 2.44 μmol/g，CH_3OH 生成速率约为 0.22 μmol/(g·h)。

与 Zn_2GeO_4 纳米棒光催化剂相比，光照 10 h，Zn_2GeO_4/ZIF-8 的 CH_3OH 产量提高了 62%。Zn_2GeO_4/ZIF-8 在光催化还原 CO_2 方面表现出比纯 Zn_2GeO_4 更高的活性，这主要归因于复合材料中 ZIF-8 组元的引入。首先，ZIF-8 能有效吸附水溶液中溶解的 CO_2；其次，Zn_2GeO_4/ZIF-8 的吸光性优于

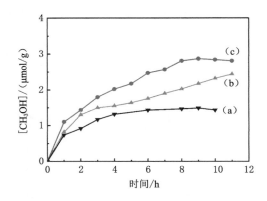

图 3-12　CH_3OH 在 Zn_2GeO_4、Zn_2GeO_4/ZIF-8 和装载 1%Pt 的 Zn_2GeO_4/ZIF-8 的生成随光照时间的变化曲线

Zn_2GeO_4；紫外-可见光谱（图 3-13）观察到，与 Zn_2GeO_4 相比，Zn_2GeO_4/ZIF-8 的光谱吸收边略有红移。Zn_2GeO_4/ZIF-8 纳米棒负载 Pt[图 3-12(c)]作为辅助催化剂可以进一步提高 CH_3OH 的生成速率。这是由于 Pt 助催化剂可以提高光生电子-空穴对的分离，这在 CO_2 的光催化还原和水的分解中得到了证明。需要强调的是，Zn_2GeO_4/ZIF-8 纳米棒的 XRD 衍射谱在光催化反应前后没有变化（图 3-14），说明了 Zn_2GeO_4/ZIF-8 纳米棒在光照射下具有很高的稳定性。

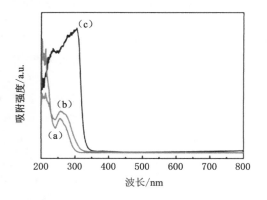

图 3-13　样品的紫外-可见吸收光谱

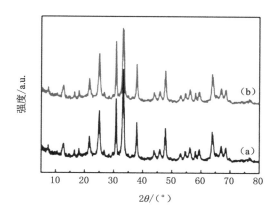

图 3-14 Zn_2GeO_4/ZIF-8 光催化反应前后的 XRD 图

3.4 本章小结

通过在 Zn_2GeO_4 纳米棒表面原位生长 ZIF-8 颗粒,成功制备了 Zn_2GeO_4/ZIF-8 杂交纳米棒。通过调节反应时间等反应参数,可以控制纳米棒上 ZIF-8 纳米颗粒的数量。Zn_2GeO_4/ZIF-8 纳米棒在水体系中比纯 Zn_2GeO_4 纳米棒表现出更高的 CO_2 光催化活性,这一方面是由于 ZIF-8 引入后,提高了 Zn_2GeO_4/ZIF-8 对 CO_2 的吸附性能;另一方面,与纯 Zn_2GeO_4 相比,Zn_2GeO_4/ZIF-8 纳米棒具有更好的光响应。本书为开发更多高活性光复合催化剂提供了一种策略,并且该方法可应用于其他催化剂体系,以提高 CO_2 光催化转化效率。

参考文献

[1] INOUE T, FUJISHIMA A, KONISHI S, et al. Photoelectrocatalytic reduction of carbon dioxide in aqueous suspensions of semiconductor powders[J]. Nature, 1979, 277(5698): 637-638.

[2] WANG W N, AN W J, RAMALINGAM B, et al. Size and structure matter: enhanced CO_2 photoreduction efficiency by size-resolved ultrafine Pt nanoparticles on TiO_2 single crystals[J]. Journal of the American

chemical society,2012,134(27):11276-11281.

[3] TU W G,ZHOU Y,LIU Q,et al.Robust hollow spheres consisting of alternating titania nanosheets and graphene nanosheets with high photocatalytic activity for CO_2 conversion into renewable fuels[J].Advanced functional materials,2012,22(6):1215-1221.

[4] RYU J,LEE S H,NAM D H,et al.Rational design and engineering of quantum-dot-sensitized TiO_2 nanotube arrays for artificial photosynthesis[J]. Advanced materials,2011,23(16):1883-1888.

[5] FUJIWARA H,HOSOKAWA H,MURAKOSHI K,et al.Effect of surface structures on photocatalytic CO_2 reduction using quantized CdS nanocrystallites[J].The journal of physical chemistry B,1997,101(41): 8270-8278.

[6] XIE K,UMEZAWA N,ZHANG N,et al.Self-doped $SrTiO_{3-\delta}$ photocatalyst with enhanced activity for artificial photosynthesis under visible light[J].Energy and environmental science,2011,4(10):4211.

[7] XI G C,OUYANG S X,LI P,et al. Ultrathin $W_{18}O_{49}$ nanowires with diameters below 1 nm: synthesis, near-infrared absorption, photoluminescence,and photochemical reduction of carbon dioxide[J].Angewandte chemie,2012,124(10):2445-2449.

[8] PENG L,ZHANG J L,LI J S,et al.Surfactant-directed assembly of mesoporous metal-organic framework nanoplates in ionic liquids[J].Chemical communications(Cambridge,England),2012,48(69):8688-8690.

[9] YAN S C,OUYANG S X,GAO J,et al. A room-temperature reactive-template route to mesoporous $ZnGa_2O_4$ with improved photocatalytic activity in reduction of CO_2 [J]. Angewandte chemie, 2010, 122(36): 6544-6548.

[10] LIU Q,ZHOU Y,KOU J H,et al.High-yield synthesis of ultralong and ultrathin Zn_2GeO_4 nanoribbons toward improved photocatalytic reduction of CO_2 into renewable hydrocarbon fuel [J]. Journal of the American chemical society,2010,132(41):14385-14387.

[11] ZHANG N,OUYANG S X,LI P,et al.Ion-exchange synthesis of a micro/mesoporous Zn_2GeO_4 photocatalyst at room temperature for photoreduction of CO_2 [J]. Chemical communications (Cambridge,

England),2011,47(7):2041-2043.

[12] LIU Q, ZHOU Y, TIAN Z P, et al. Zn_2GeO_4 crystal splitting toward sheaf-like, hyperbranched nanostructures and photocatalytic reduction of CO_2 into CH_4 under visible light after nitridation[J]. Journal of materials chemistry,2012,22(5):2033-2038.

[13] HOFFMANN M R, MOSS J A, BAUM M M. Artificial photosynthesis: semiconductor photocatalytic fixation of CO_2 to afford higher organic compounds[J]. Dalton transactions (Cambridge, England), 2011, 40(19):5151-5158.

[14] TAN Y N, WONG C L, MOHAMED A R. An overview on the photocatalytic activity of nano-doped-TiO_2 in the degradation of organic pollutants[J]. ISRN materials science,2011,2011:261219.

[15] INDRAKANTI V P, KUBICKI J D, SCHOBERT H H. Photoinduced activation of CO_2 on Ti-based heterogeneous catalysts: current state, chemical physics-based insights and outlook[J]. Energy and environmental science,2009,2(7):745.

[16] TONG H, OUYANG S X, BI Y P, et al. Nano-photocatalytic materials: possibilities and challenges[J]. Advanced materials, 2012, 24(2): 229-251.

[17] ANPO M, YAMASHITA H, IKEUE K, et al. Photocatalytic reduction of CO_2 with H_2O on Ti-MCM-41 and Ti-MCM-48 mesoporous zeolite catalysts[J]. Catalysis today,1998,44(1/2/3/4):327-332.

[18] IKEUE K, YAMASHITA H, ANPO M, et al. Photocatalytic reduction of CO_2 with H_2O on Ti-β zeolite photocatalysts: effect of the hydrophobic and hydrophilic properties[J]. The journal of physical chemistry B, 2001,105(35):8350-8355.

[19] YANG C C, VERNIMMEN J, MEYNEN V, et al. Mechanistic study of hydrocarbon formation in photocatalytic CO_2 reduction over Ti-SBA-15 [J]. Journal of catalysis,2011,284(1):1-8.

[20] PARK K S, NI Z, CÔTÉ A P, et al. Exceptional chemical and thermal stability of zeolitic imidazolate frameworks[J]. Proceedings of the national academy of sciences of the United States of America, 2006, 103 (27):10186-10191.

[21] WANG B, CÔTÉ A P, FURUKAWA H, et al. Colossal cages in zeolitic imidazolate frameworks as selective carbon dioxide reservoirs[J]. Nature, 2008, 453(7192):207-211.

[22] BANERJEE R, PHAN A, WANG B, et al. High-throughput synthesis of zeolitic imidazolate frameworks and application to CO_2 capture[J]. Science, 2008, 319(5865):939-943.

[23] WANG C, XIE Z G, DEKRAFFT K E, et al. Doping metal-organic frameworks for water oxidation, carbon dioxide reduction, and organic photocatalysis[J]. Journal of the American chemical society, 2011, 133(34):13445-13454.

[24] HORIUCHI Y, TOYAO T, SAITO M, et al. Visible-light-promoted photocatalytic hydrogen production by using an amino-functionalized Ti(IV) metal-organic framework[J]. The journal of physical chemistry C, 2012, 116(39):20848-20853.

[25] ISIMJAN T T, KAZEMIAN H, ROHANI S, et al. Photocatalytic activities of Pt/ZIF-8 loaded highly ordered TiO_2 nanotubes[J]. Journal of materials chemistry, 2010, 20(45):10241.

[26] FU Y H, SUN D R, CHEN Y J, et al. An amine-functionalized titanium metal-organic framework photocatalyst with visible-light-induced activity for CO_2 reduction[J]. Angewandte chemie, 2012, 124(14):3420-3423.

[27] ZHANG L, CAO X F, MA Y L, et al. Microwave-assisted preparation and photocatalytic properties of Zn_2GeO_4 nanorod bundles[J]. CrystEngComm, 2010, 12(10):3201.

[28] YAN S C, WAN L J, LI Z S, et al. Facile temperature-controlled synthesis of hexagonal Zn_2GeO_4 nanorods with different aspect ratios toward improved photocatalytic activity for overall water splitting and photoreduction of CO_2[J]. Chemical communications (Cambridge, England), 2011, 47(19):5632-5634.

[29] MA B J, WEN F Y, JIANG H F, et al. The synergistic effects of two cocatalysts on Zn_2GeO_4 on photocatalytic water splitting[J]. Catalysis letters, 2010, 134(1/2):78-86.

[30] D'ALESSANDRO D, SMIT B, LONG J. Carbon dioxide capture: pros-

pects for new materials[J]. Angewandte chemie international edition, 2010, 49(35): 6058-6082.

[31] CRAVILLON J, MÜNZER S, LOHMEIER S J, et al. Rapid room-temperature synthesis and characterization of nanocrystals of a prototypical zeolitic imidazolate framework[J]. Chemistry of materials, 2009, 21(8): 1410-1412.

[32] BAMWENDA G, TSUBOTA S, NAKAMURA T, et al. The influence of the preparation methods on the catalytic activity of platinum and gold supported on TiO_2 for CO oxidation[J]. Catalysis letters, 1997, 44: 83-87.

[33] LIU Q, ZHOU Y, MA Y, et al. Synthesis of highly crystalline $In_2Ge_2O_7$(En) hybrid sub-nanowires with ultraviolet photoluminescence emissions and their selective photocatalytic reduction of CO_2 into renewable fuel[J]. RSC advances, 2012, 2(8): 3247.

[34] MCCARTHY M C, VARELA-GUERRERO V, BARNETT G V, et al. Synthesis of zeolitic imidazolate framework films and membranes with controlled microstructures[J]. Langmuir, 2010, 26(18): 14636-14641.

[35] HUANG A S, BUX H, STEINBACH F, et al. Molecular-sieve membrane with hydrogen permselectivity: ZIF-22 in LTA topology prepared with 3-aminopropyltriethoxysilane as covalent linker[J]. Angewandte chemie international edition, 2010, 49(29): 4958-4961.

[36] ROY S C, VARGHESE O K, PAULOSE M, et al. Toward solar fuels: photocatalytic conversion of carbon dioxide to hydrocarbons[J]. ACS nano, 2010, 4(3): 1259-1278.

[37] SATO J, KOBAYASHI H, IKARASHI K, et al. Photocatalytic activity for water decomposition of RuO_2-dispersed Zn_2GeO_4 with d^{10} configuration[J]. Cheminform, 2004, 35(27): 27016.

[38] HUANG J H, WANG X C, HOU Y D, et al. Degradation of benzene over a zinc germanate photocatalyst under ambient conditions[J]. Environmental science and technology, 2008, 42(19): 7387-7391.

第4章 多孔 ZnO 纳米片薄膜的制备及其在染料敏化太阳能电池中的应用研究

4.1 引言

类沸石咪唑酯骨架材料(ZIFs)是一类具有类沸石结构的金属有机框架(MOFs),由于其优异的热稳定性和化学稳定性,在气体吸附、分子分离和催化等方面表现出巨大的应用潜力[1-3]。到目前为止,使用不同的咪唑配体应用于不同的有机溶剂,如二甲基甲酰胺(DMF)、二乙基甲酰胺(DEF)和甲醇[4-6],设计和合成了大量的 ZIFs。在目前合成出的 ZIFs 中,只有 ZIF-8 和 ZIF-67 等少数几种类型的 ZIFs 可以在水溶液中合成[7-9]。本课题组以六水合硝酸锌和 2-甲基咪唑在常温去离子水中合成了一种新的二维 ZIF 结构 (ZIF-L)[10]。ZIF-L 呈树叶状形态,在层间有一个尺寸为 9.4 Å×7.0 Å×5.3 Å 的空腔,表现出良好的 CO_2 吸附性能。在实际应用中,常温水中原位生长合成方法将大大降低 ZIFs 大规模生产的成本和环境影响。

近年来,学者们对于以 ZIFs(MOFs)作为模板/前驱体制备的衍生材料研究愈加深入,不仅使 ZIFs(MOFs)得到快速发展,还加快了其投入实际应用的脚步。ZIFs(MOFs)研究如此火热的原因,一方面是以 MOFs 为模板/前驱体所制备得到的衍生材料与传统多孔纳米材料相比而言具有更多的特点和优势,如可调控的结构与性质、独特的孔道结构、功能多样化和制备工艺简单等;另一个方面就是所得到的衍生材料通常具有较大的比表面积、较高的孔隙率和高度分散的活性位点。同时,MOF 结构中金属离子或金属团簇与有机配体组成的周期性网状结构在热解过程中能够有效防止金属纳米粒子或金属氧化物等纳米结构的聚集。因此,以 ZIFs(MOFs)作为模板/前驱体制备了大量衍生材料,包括金属硫化物或氧化物,如 ZnS[11]、Fe_2O_3[12-13]、MgO[14]、

Co_3O_4[15-16]、$GdCoO_3$[17]、Fe_2O_3/TiO_2纳米复合材料[18]以及金属/金属氧化物纳米晶[19];碳基材料,如多孔碳[20-21]、$C-Fe_3O_4$[22]和碳包覆ZnO量子点[23]等。虽然已经可以在不同的衬底上制备出ZIFs薄膜,并将其应用到气体分离中,但目前为止还没有直接由ZIFs薄膜衍生出金属氧化物薄膜的报道。

本章采用原位生长法,室温下水溶液中在FTO玻璃沉积了ZIF-L膜,将ZIF-L膜在空气中煅烧后直接转化为多孔的ZnO膜,合成方法简单,不需要对FTO玻璃进行表面改性,制备所得的薄膜由多孔ZnO纳米片组成,纳米片厚度约4 μm,可作为染料敏化太阳能电池(DSSC)的光阳极。由于ZnO薄膜独特的片状多孔结构,DSSC光转换效率达到2.52%,是ZnO纳米棒电池的2倍(1.27%)。

4.2 实验部分

4.2.1 ZnO纳米片薄膜的合成

所有化学制品均为分析纯,使用时无须进一步纯化。将$F:SnO_2$(FTO)导电玻璃衬底用洗涤剂、稀盐酸水溶液、KOH乙醇溶液和去离子水在超声波清洗机中清洗,最后用氮气吹干备用。采用ZIF-L前驱体模板法制备了氧化锌纳米片。树叶状ZIF-L粉体的制备在以前的文章[10]中已经详细报道过。一般分别取0.587 g $Zn(NO_3)_2 \cdot 6H_2O$和1.298 g 2-甲基咪唑(Hmim)分别溶于40 mL去离子水中,磁力搅拌下将Hmim溶液倒入$Zn(NO_3)_2 \cdot 6H_2O$溶液中。室温下搅拌2 h,离心,将固体产物从乳状胶体中分离出来,得到的白色ZIF-L粉用去离子水冲洗,离心3次,60 ℃干燥。

将清洗后的FTO玻璃基板室温下垂直放置在装有新鲜ZIF-L合成溶液的烧杯中2 h。如果需要,可以重复这种ZIF-L薄膜的生长。所得到的薄膜用去离子水清洗,并在60 ℃的烤箱中干燥。最后,在空气中以1 ℃/min的加热速率煅烧0.5 h,得到多孔ZnO薄膜。

4.2.2 薄膜表征

用粉末X射线衍射(XRD)(Rigaku Ultima Ⅲ,CuKa射线)测定了制备产物的晶相结构。样品的形貌通过扫描电子显微镜(FEI Nova NanoSEM 450 FEG SEM)和透射电子显微镜(JEOL 3010显微镜)观察。用X射线光电子能谱(XPS,K-Alpha,Thermo Fishers Scientific)分析了薄膜的化学成分。采用气体吸附分析仪（Micromertics ASAP 2020）在 77 K 进行 N_2 吸附-脱附实验,测定样品的 BET 表面积和孔径分布。采用 Netzsch STA449F1 同步热分析仪同时进行样品的热重（TG）和差热（DTA）测试。将约 6 mg 样品装入氧化铝坩埚中,在空气中以 5 ℃/min 的升温速率从室温加热至 700 ℃。利用衰减全反射（ATR）技术,在 BRUKER Tensor 27 红外光谱仪上测试了样品的傅里叶变换红外光谱（FTIR）。用紫外-可见分光光度计（UV-2550,Shimadzu）在室温下测试样品的紫外-可见漫反射光谱。

4.2.3 太阳能电池制备

在 F-SnO_2（FTO）导电玻璃（4 mm,10 Ω/m^2）电极上生长厚度为 4~5 μm（4 mm×4 mm）的 ZIF-L 或 ZnO 纳米棒薄膜。制备的 ZIF-L 薄膜在 60 ℃ 空气中放置 12 h 干燥后,在 550 ℃ 空气中退火 0.5 h。当薄膜冷却至 80 ℃ 后,将薄膜浸泡到乙腈-甲苯（1∶1）混合溶液中 3 h。采用热分解法在 FTO 衬底上制备了 Pt 对电极。工作电极和对电极用 25 μm 厚的 Surlyn 树脂连接在一起。采用真空填充法将 50 mmol/L LiI、1.0 mol/L 1,2-二甲基-3-丙基咪唑碘化物、30 mmol/L I_2、0.10 mol/L 硫氰酸胍和 50 mmol/L 叔丁基吡啶的电解液填充到两个电极之间。电解注入孔用热熔胶密封,两孔用载玻片封住。

4.2.4 光伏特性测试

制备的太阳能电池用黑色金属掩模板盖住,露出 0.36 cm^2 面积进行测试。通过 Keithley 2400 数字源表测定电池在 AM1.5 模拟太阳光下的 J-V 曲线。在 Keithley 2400 数字源表上,采用 Oriel-cornertm-2601/4m 单色仪,在 300 W 氙灯照射下,测试了入射光子电流转换效率（IPCE）谱。利用太阳模拟器和电化学工作站测量了其瞬态电流。在 AM1.5 太阳辐射下,在 50 mHz~100 kHz 的恒电位仪上,对 DSSCs 的电化学阻抗谱（EIS）曲线进行了研究。

4.3 结果与讨论

4.3.1 ZIF-L 薄膜的结构表征

在不搅拌的情况下,将清洁的 FTO 玻璃衬底垂直放置在六水合硝酸锌和 2-甲基咪唑（Hmin）的水溶液中,能够实现 ZIF-L 薄膜的大面积制备。ZIF-L 的组成通过元素分析确定为 $C_{10}H_{16}N_5O_{3/2}Zn$,如之前的工作所示[10]。ZIF-L 晶体由不同 Zn 离子形成的六边形和平行四边形相互连接,在 ab 平面上形成二维层网,然后沿着 c 方向堆叠。

采用场发射扫描电子显微镜(FE-SEM)对室温下沉积 4 h 的样品进行表征,如图 4-1(a)、(b)和图 4-2 所示。ZIF-L 薄膜是由高度均匀且无裂纹的纳米片组成[图 4-1(a)和图 4-2]。由高倍放大的 SEM 图像[图 4-1(b)]和[图 4-2(c)、(d)]可以看出,纳米片表面相互交错生长,没有破裂,每片纳米片的厚度为 100～150 nm。FE-SEM 断面图[图 4-1(c)]进一步证实了连续膜具有纳米片结构,大部分纳米片垂直生长在基底上,膜厚约为 4 μm。用粉末 X 射线衍射(XRD)检测了薄膜和粉末的结晶结构和相纯度。ZIF-L 薄膜和粉末的 XRD 衍射图谱如图 4-1(d)所示。ZIF-L 粉体的 XRD 图谱与我们之前的工作[10]中报道的图谱吻合良好,证实了 ZIF-L 的合成成功。XRD 结果还表明,ZIF-L 薄膜具有较强的(112)取向,这与 SEM 观察结果一致。

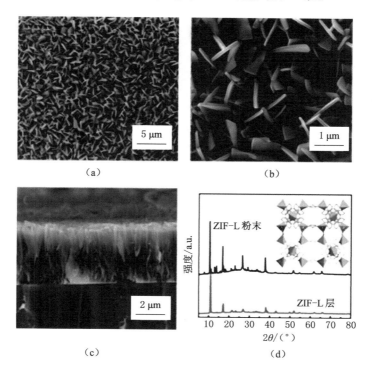

图 4-1 ZIF-L 薄膜的 FE-SEM 图

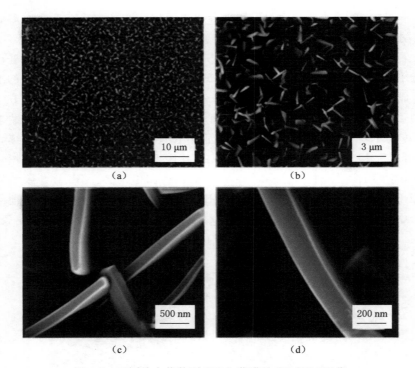

图 4-2 不同放大倍数下 ZIF-L 薄膜的 FE-SEM 图像

对不同沉积时间制备的样品采用 FE-SEM 进行表征,研究了 ZIF-L 薄膜的形成过程,其结果如图 4-3 所示。沉积 1 h 样品的 FE-SEM 图像清楚地显示,在结晶早期,FTO 玻璃上垂直生长出一些树叶状纳米片,纳米片的尺寸不均匀,大小相差比较大[图 4-3(a)]。有一些像叶芽一样的小纳米片,也有一些像叶子一样的大纳米片。FTO 上的这些叶状颗粒与在溶液中生长的 ZIF-L 粉末具有相似的形态。通过延长反应时间到 2 h,一些小叶芽长成大的纳米薄片,纳米片尺寸明显增加。反应 4 h 后,所有的小芽状纳米片大都以 ZIF-L 晶体结构的 a、b 方向向上生长,并在 c 方向相互连接,形成连续的 ZIF-L 膜。将反应时间进一步延长至 6 h 后,纳米片厚度显著增加至 400~500 nm[图 4-3(e)、(f)]。

4.3.2 ZnO 薄膜的结构表征

通过空气中热处理可以将制备的 ZIF-L 薄膜原位转化为多孔氧化锌纳米片。采用热重分析(TGA)研究了 ZIF-L 的热稳定性,结果如图 4-4 所示。ZIF-L

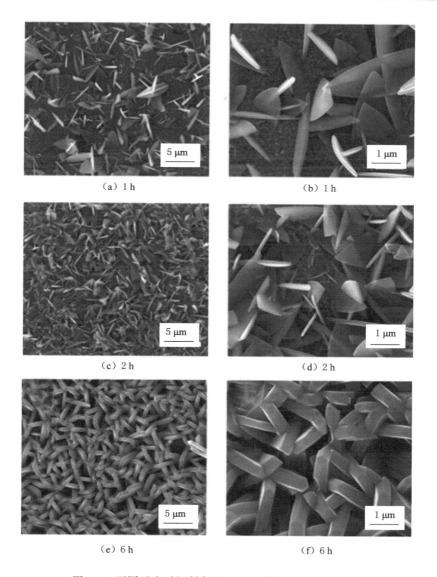

图 4-3　不同反应时间制备的 ZIF-L 薄膜的 FE-SEM 图

薄膜在 280 ℃时呈现出 12.2%的失重,与之对应的是纳米薄片表面的水分子的去除和未反应配体(如 2-甲基咪唑)的解吸附。在 280~550 ℃的温度范围内,ZIF-L 膜出现显著的失重,高达 47.7%,表明在这个温度区间,ZIF-L 氧化分解并形成了 ZnO。所以确定 ZIF-L 薄膜热处理转化为 ZnO 的温度定为 550 ℃。

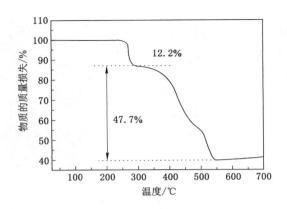

图 4-4 ZIF-L 膜在空气中的热重曲线

图 4-5 是 ZIF-L 薄膜 550 ℃煅烧前后的红外光谱(FTIR)分析结果图。由 ZIF-L 薄膜的 FTIR 峰曲线可知,423 cm^{-1} 对应 Zn-N 拉伸峰,1 146 cm^{-1} 和 1 307 cm^{-1} 为—CH 振动峰,1 384 cm^{-1} 对应的是 C—C 拉伸峰,1 565 cm^{-1} 为 C=N 拉伸峰,煅烧以后这些特征峰都消失了。说明 ZIF-L 薄膜在经过 550 ℃煅烧后已经完全转变成了一种无机材料。

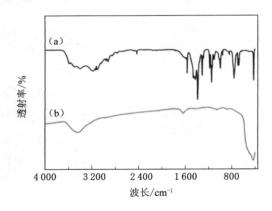

图 4-5 样品的 FTIR 谱

X 射线光电子能谱(XPS)测试结果如图 4-6 所示,根据测试结果可以知道,N 1s 在 ZIF-L 薄膜中的位置约为 398.58 eV,对应 C—N 键中的氮和 2-甲基咪唑中的氮[24]。煅烧后样品中没有 N 峰,说明在经过 550 ℃煅烧后条件下

ZIF-L 薄膜已经完全分解。

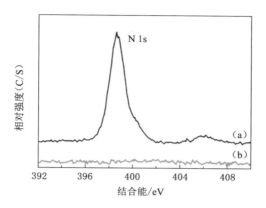

图 4-6　XPS N 1s 光谱

在经过 550 ℃煅烧后 ZIF-L 薄膜的 XRD 衍射图如图 4-7 所示。在衍射角 $2\theta=31.90°、34.39°、36.29°、47.60°、56.70°、62.89°$ 处都出现了 ZnO 的衍射峰，对应(100)、(002)、(101)、(102)、(110)、(103)晶面样品的峰位与 JCPDS 卡标准值(JCPDS36-1451)基本符合。X 射线衍射图表明煅烧样品具有六角纤锌矿晶体结构，即 ZIF-L 薄膜在 550 ℃的空气中转变为纯 ZnO 相。

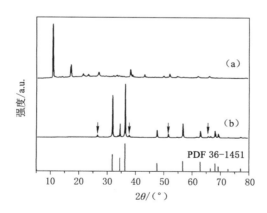

图 4-7　样品的 XRD 图

煅烧后纳米片的形态和厚度保持不变，纳米片呈花瓣状弯曲(图 4-8)。在煅烧过程中，ZIF-L 薄膜的有机配体被氧化，释放气体，从而在纳米片表面留下规则均匀的多孔结构。制备的氧化锌纳米片表面粗糙,由纳米颗粒组成。纳米片中形成了大量的纳米孔,这是由于 ZIF-L 分解导致热处理过程中挥发性气体如 H_2O、N_xO 和 CO_2 的释放。由图 4-8(b)可知,纳米片的孔尺寸大小为 20 nm×150 nm。此外,高分辨的 FE-SEM 图[图 4-8(c)]和 ZnO 薄膜单个纳米片的 TEM 图[图 4-8(d)]清楚地表明,纳米片是由相互连接的纳米颗粒组成的。但是纳米粒子的粒径分布很广,从 20 nm 到 150 nm 不等。ZnO 纳米片的晶格条纹清晰且相互平行,相邻晶格平面间的平面间距为 2.60 Å,对应于六角纤锌矿 ZnO 的(002)平面。然而,与 ZIF-L 膜不同的是,采用 ZIF-L 粉末在 550 ℃下煅烧 0.5 h 后,无法保持叶片状的形貌,得到的 ZnO 纳米片比 ZIF-L 膜生成的致密得多(图 4-9)。

图 4-8　ZnO 膜的 FE-SEM 图像

用 ZIF-L 薄膜制备的样品记为 ZnO NR,用 ZIF-L 粉末煅烧制备记为 ZnO PS。ZIF-L 薄膜(ZnO NS)和 ZIF-L 粉末的 BET 表面积分别约为 69 m^2/g 和 5 m^2/g。氮吸附测量结果表明,ZnO NS 具有双峰孔径分布

第 4 章 多孔 ZnO 纳米片薄膜的制备及其在染料敏化太阳能电池中的应用研究

图 4-9 在 550 ℃下 ZIF-L 粉末煅烧样品的 FE-SEM 图像

(图 4-10);11 nm 的气孔可能是由于热处理过程中气体的释放而产生的,而 76 nm 的气孔则是由于相邻纳米片之间的开放空间造成的。同时,一些细小的孔洞可能会相互连接,形成较大的孔洞,尺寸分布较广。ZnO PS 样品没有明显的孔结构。

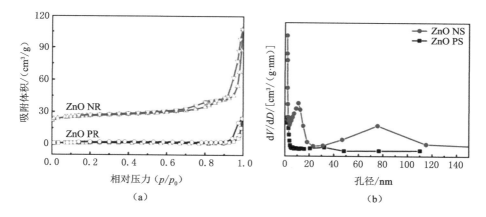

图 4-10 ZnO 样品的 N_2 吸附脱附等温线和两个样品对应的孔径分布

4.3.3 ZnO 基 DSSCs 的表征

众所周知,ZnO 是最常见的半导体材料之一,由于其较高的电子迁移率和合适的能带结构,在光催化降解有机污染物和染料敏化太阳能电池(DSSCs)方面具有优异的性能。氧化锌基材料或器件的性能在很大程度上取决于其几何和形貌结构特征,尤其是 ZnO 基 DSSC。但 ZnO 纳米颗粒在煅烧过程中难以保持光阳极的多孔结构。由纳米颗粒制备光阳极的过程也很复杂,常常导致较差的重复性。因此,半导体薄膜在衬底上的直接生长作为 DSSCs 受到了特别的关注。

本部分研究了 FTO 玻璃上的 ZIF-L 膜制备的多孔 ZnO 纳米片膜(ZnO NR)作为 DSSC 光阳极的潜力,并与 ZnO 纳米棒(ZnO NR)基光阳极的性能进行了比较。ZnO NR 膜按照文献报道的方法制备[25-27]。不同放大倍数的 FE-SEM 图像证实,ZnO NR 薄膜确实均匀致密地生长在 FTO 衬底上,薄膜厚度也在 4 μm 左右(图 4-11)。

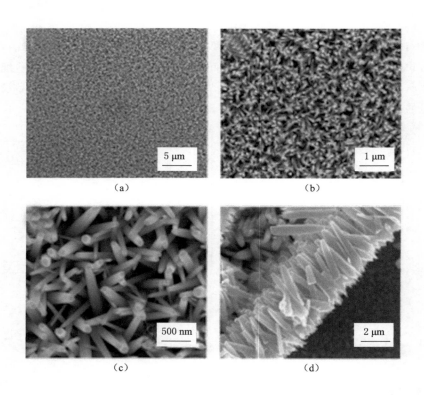

图 4-11　ZnO NR 薄膜不同放大率下的 FE-SEM 图像

图 4-12 为两种太阳能电池在暗态和 AM 1.5 全光照(100 mW/cm^2)下的光电流-电压(J-V)特性对比图,两种电池的暗电流接近于零。对于 ZnO NS

基电池,J-V 曲线显示短路电流密度(Jsc)为 6.87 mA/cm², 开路电压(Voc)为 565 mV, 填充系数(FF)为 64.9%, 转换效率(η)为 2.52%。相比之下, ZnO NR 基电池的 Jsc 为 3.74 mA/cm², Voc 为 559 mV, FF 为 60.7%, 而 η 为 1.27%。

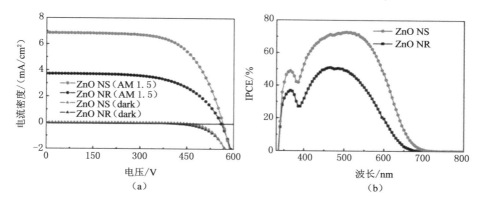

图 4-12 电池 J-V 曲线及对应的 IPCE 光谱

基于氧化锌纳米棒的 DSSC 电池的 η 与文献报道的 ITO 玻璃衬底氧化锌纳米线纳米棒电池很接近[28-31],远高于类似氧化锌薄膜厚度的电池效率(0.83%)[32](表 4-1)。对于这两种 ZnO DSSCs,当打开光时,光电流迅速达到最大值并在之后保持恒定[图 4-13(a)],这表明不存在质量传输问题。

表 4-1 不同 ZnO 膜的 DSSC 太阳能电池的特性

光阳极结构	ZnO 厚/μm	η/%	Jsc/(mA/cm²)	Voc/mV	FF	Ref
ZnO 纳米线	7	0.45	1.52	636	0.48	28(2011 年)
ZnO 纳米线	几十	2.1	—	—	—	29(2013 年)
ZnO 纳米棒	5.1	1.18	3.24	680	0.58	31(2010 年)
ZnO 纳米线	5.5	0.84	3.4	500	0.49	32(2007 年)

ZnO NS 基电池的转换效率 η 比 ZnO NR 基电池的高 100%。仔细比较参数可以发现,这两种 DSSC 的 Voc 值几乎相同,这仅仅取决于 ZnO 的平带电位(V_{fb})和 I^-/I_3 的氧化还原电位。Jsc 有明显的提高,这是由光的捕获效率、电荷注入效率和电子的收集决定的。纳米片和纳米棒膜吸附的 MK2 染料量分别为 3.2×10^{-7} mol/cm² 和 2.1×10^{-7} mol/cm²,这可能是由于这两种

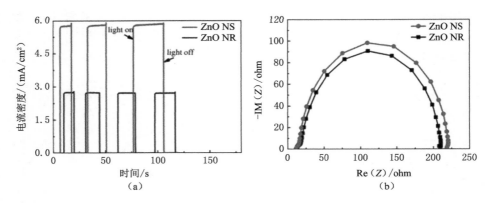

图 4-13　DSSCs 电池的瞬态电流曲线与 DSSCs 的阻抗谱

膜的比表面积不同导致的。

图 4-12(b)显示了入射单色光子对 DSSC 电流转换效率(IPCE)光谱的波长分布。在整个 350～700 nm 的光谱范围内，ZnO NS 基电池的 IPCE 远远高于 ZnO NR 基电池，这与 Jsc 值很好地吻合。在 505 nm 和 465 nm 波长下，ZnO NS 电池的最大 IPCE 分别为 72.7% 和 51%。ZnO NS 膜的 IPCE 比 ZnO NR 膜的 IPCE 高出 40% 以上。然而，有报道称，当染料吸收增加 1.2 倍时，IPCE 仅会增加约 3%[33-34]。

在 AM 1.5 太阳辐射下，在 50 mHz～100 kHz 的恒电位仪上，对 DSSCs 的电化学阻抗谱(EIS)曲线进行了研究。由图 4-13(b)分析可知，ZnO NR 基电池的阻抗略低于 ZnO NS 基电池，这是由于其一维纳米结构，电子(空穴)产生和传输速度更快，空穴-电子复合速率更低[35-38]。

较高的 IPCE 值也可以归因于薄膜的光散射能力增强，促进了 MK2 染料的光捕获。UV-vis 漫反射光谱可以证实这一点(图 4-14)。相对于 ZnO NR 膜，ZnO NS 膜在可见光和近红外区域具有更高的漫反射能力，说明入射光在膜内具有明显的散射。

现在可以清楚地得出结论，ZnO 膜的结构对相关 DSSC 的性能起着关键作用：

① 高比表面积的 ZnO 膜导致更高的染料负载。

② 每个 ZnO-NS 中的纳米孔能够更有效地进行电解液扩散和快速吸附染料，从而减少 Zn^{2+}/染料的聚集。

③ 这种微米级的基于纳米片的架构通过散射增强和捕获(光染料相互作用)提高了光捕获效率。

第4章 多孔ZnO纳米片薄膜的制备及其在染料敏化太阳能电池中的应用研究

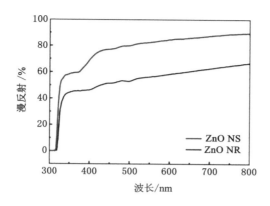

图4-14 ZnO NS和ZnO NR薄膜的漫反射光谱

4.4 本章小结

室温水溶液中在FTO玻璃上生长了ZIF-L纳米片薄膜,而且无须对基底进行表面改性。将ZIF-L膜在550 ℃的空气中直接煅烧0.5 h,成功制备了具有独特纳米片结构的多孔ZnO膜。多孔ZnO膜作为染料敏化太阳能电池的光阳极,整体光转换效率为2.52%,是制备的相同厚度ZnO纳米棒(1.27%)光转换效率的2倍。ZnO-NS薄膜的独特结构对相关DSSC的性能提升起着关键作用。本书为制备多孔氧化锌纳米片膜提供了一种新的有效方法,这种制备方法一般适用于直接合成金属氧化物等具有某些其他特殊性能的材料,如金属、金属氧化物或金属沉积氧化物、碳/金属氧化物,C和N掺杂金属氧化物。ZIF(MOFs)等材料目前在光电子材料、太阳能电池、催化剂、超级电容器和传感器等方面有很大的应用前景。

参考文献

[1] BANERJEE R, PHAN A, WANG B, et al. High-throughput synthesis of zeolitic imidazolate frameworks and application to CO_2 capture[J]. Science, 2008, 319(5865):939-943.

[2] WANG B, CÔTÉ A P, FURUKAWA H, et al. Colossal cages in zeolitic imidazolate frameworks as selective carbon dioxide reservoirs[J].

Nature,2008,453(7192):207-211.

[3] PARK K S,NI Z,CÔTÉ A P,et al.Exceptional chemical and thermal stability of zeolitic imidazolate frameworks[J].Proceedings of the national academy of sciences of the United States of America,2006,103(27):10186-10191.

[4] BUX H,LIANG F Y,LI Y S,et al.Zeolitic imidazolate framework membrane with molecular sieving properties by microwave-assisted solvothermal synthesis[J].Journal of the American chemical society,2009,131(44):16000-16001.

[5] CRAVILLON J,MÜNZER S,LOHMEIER S J,et al.Rapid room-temperature synthesis and characterization of nanocrystals of a prototypical zeolitic imidazolate framework[J].Chemistry of materials,2009,21(8):1410-1412.

[6] CORMA A,GARCÍA H,XAMENA F X L I.Engineering metal organic frameworks for heterogeneous catalysis[J].Chemical reviews,2010,110(8):4606-4655.

[7] PAN Y C,LIU Y Y,ZENG G F,et al.Rapid synthesis of zeolitic imidazolate framework-8 (ZIF-8) nanocrystals in an aqueous system [J].Chemical communications(Cambridge,England),2011,47(7):2071-2073.

[8] KIDA K,OKITA M,FUJITA K,et al.Formation of high crystalline ZIF-8 in an aqueous solution[J].CrystEngComm,2013,15(9):1794.

[9] YAO J F,HE M,WANG K,et al.High-yield synthesis of zeolitic imidazolate frameworks from stoichiometric metal and ligand precursor aqueous solutions at room temperature[J].CrystEngComm,2013,15(18):3601.

[10] CHEN R Z,YAO J F,GU Q F,et al.A two-dimensional zeolitic imidazolate framework with a cushion-shaped cavity for CO_2 adsorption[J].Chemical communications(Cambridge,England),2013,49(82):9500-9502.

[11] JIANG Z,SUN H,QIN Z,et al.Synthesis of novel ZnS nanocages utilizing ZIF-8 polyhedral template[J].Chemical communications(Cambridge,England),2012,48(30):3620-3622.

[12] ZHANG L,WU H B,MADHAVI S,et al.Formation of Fe_2O_3 microboxes with hierarchical shell structures from metal-organic frameworks and their lithium storage properties[J].Journal of the American chemical society,2012,134(42):17388-17391.

[13] XU X D,CAO R G,JEONG S,et al.Spindle-like mesoporous α-Fe_2O_3 anode material prepared from MOF template for high-rate lithium batteries[J].Nano letters,2012,12(9):4988-4991.

[14] KIM T K,LEE K J,CHEON J Y,et al.Nanoporous metal oxides with tunable and nanocrystalline frameworks via conversion of metal-organic frameworks[J].Journal of the American chemical society,2013,135(24):8940-8946.

[15] WANG W X,LI Y W,ZHANG R J,et al.Metal-organic framework as a host for synthesis of nanoscale Co_3O_4 as an active catalyst for CO oxidation[J].Catalysis communications,2011,12(10):875-879.

[16] MENG F L,FANG Z G,LI Z X,et al.Porous Co_3O_4 materials prepared by solid-state thermolysis of a novel Co-MOF crystal and their superior energy storage performances for supercapacitors[J].Journal of materials chemistry A,2013,1(24):7235.

[17] MAHATA P,AARTHI T,MADRAS G,et al.Photocatalytic degradation of dyes and organics with nanosized $GdCoO_3$[J].The journal of physical chemistry C,2007,111(4):1665-1674.

[18] DEKRAFFT K E,WANG C,LIN W B.Metal-organic framework templated synthesis of Fe_2O_3/TiO_2 nanocomposite for hydrogen production[J].Advanced materials,2012,24(15):2014-2018.

[19] DAS R,PACHFULE P,BANERJEE R,et al.Metal and metal oxide nanoparticle synthesis from metal organic frameworks(MOFs):finding the border of metal and metal oxides[J].Nanoscale,2012,4(2):591-599.

[20] YANG S J,KIM T,IM J H,et al.MOF-derived hierarchically porous carbon with exceptional porosity and hydrogen storage capacity[J].Chemistry of materials,2012,24(3):464-470.

[21] CHAIKITTISILP W,ARIGA K,YAMAUCHI Y.A new family of carbon materials:synthesis of MOF-derived nanoporous carbons and their promising applications[J].Journal of materials chemistry A,2013,1(1):14-19.

[22] BANERJEE A,GOKHALE R,BHATNAGAR S,et al.MOF derived porous carbon-Fe_3O_4 nanocomposite as a high performance, recyclable environmental superadsorbent[J].Journal of materials chemistry,2012,

22(37):19694.

[23] YANG S J, NAM S, KIM T, et al. Preparation and exceptional lithium anodic performance of porous carbon-coated ZnO quantum dots derived from a metal-organic framework[J]. Journal of the American chemical society, 2013, 135(20):7394-7397.

[24] MCCARTHY M C, VARELA-GUERRERO V, BARNETT G V, et al. Synthesis of zeolitic imidazolate framework films and membranes with controlled microstructures[J]. Langmuir, 2010, 26(18):14636-14641.

[25] VAYSSIERES L. Growth of arrayed nanorods and nanowires of ZnO from aqueous solutions[J]. Advanced materials, 2003, 15(5):464-466.

[26] GREENE L E, LAW M, TAN D H, et al. General route to vertical ZnO nanowire arrays using textured ZnO seeds[J]. Nano letters, 2005, 5(7):1231-1236.

[27] DAI H, ZHOU Y, CHEN L, et al. Porous ZnO nanosheet arrays constructed on weaved metal wire for flexible dye-sensitized solar cells[J]. Nanoscale, 2013, 5(11):5102-5108.

[28] KO S H, LEE D, KANG H W, et al. Nanoforest of hydrothermally grown hierarchical ZnO nanowires for a high efficiency dye-sensitized solar cell[J]. Nano letters, 2011, 11(2):666-671.

[29] KILIÇ B, WANG L Z, OZDEMIR O, et al. One-dimensional (1D) ZnO nanowires dye sensitized solar cell[J]. Journal of nanoscience and nanotechnology, 2013, 13(1):333-338.

[30] LIN C Y, LAI Y H, CHEN H W, et al. Highly efficient dye-sensitized solar cell with a ZnO nanosheet-based photoanode[J]. Energy and environmental science, 2011, 4(9):3448.

[31] KU C H, WU J J. Electron transport properties in ZnO nanowire array/nanoparticle composite dye-sensitized solar cells[J]. Applied physics letters, 2007, 91(9):093117.

[32] XIANG W C, FANG Y Y, LIN Y, et al. Polymer-metal complex as gel electrolyte for quasi-solid-state dye-sensitized solar cells[J]. Electrochimica acta, 2011, 56(3):1605-1610.

[33] NELSON J J, AMICK T J, ELLIOTT C M. Mass transport of polypyridyl cobalt complexes in dye-sensitized solar cells with mesoporous

TiO$_2$ photoanodes[J]. The journal of physical chemistry C, 2008, 112(46):18255-18263.

[34] BAXTER J B, AYDIL E S. Nanowire-based dye-sensitized solar cells[J]. Applied physics letters, 2005, 86(5):053114.

[35] LAW M, GREENE L E, JOHNSON J C, et al. Nanowire dye-sensitized solar cells[J]. Nature materials, 2005, 4(6):455-459.

[36] DAI H, ZHOU Y, LIU Q, et al. Controllable growth of dendritic ZnO nanowire arrays on a stainless steel mesh towards the fabrication of large area, flexible dye-sensitized solar cells[J]. Nanoscale, 2012, 4(17):5454-5460.

[37] MA T L, AKIYAMA M, ABE E, et al. High-efficiency dye-sensitized solar cell based on a nitrogen-doped nanostructured titania electrode[J]. Nano letters, 2005, 5(12):2543-2547.

[38] LÜ X J, MOU X L, WU J J, et al. Improved-performance dye-sensitized solar cells using Nb-doped TiO$_2$ electrodes: efficient electron injection and transfer[J]. Advanced functional materials, 2010, 20(3):509-515.

第 5 章　Fe^{3+} 掺杂 TiO_2 八面体的制备及其光催化还原 CO_2 研究

5.1　引言

自 1972 年藤岛和本田发现 TiO_2 电极上的水分解现象以来，TiO_2 的光催化性能受到了广泛的关注[1-2]。与其他半导体光催化材料相比，TiO_2 具有化学和生物稳定性高、氧化能力强、成本低、无毒等优点[3-5]。而纯锐钛矿型 TiO_2 只对紫外光有响应（λ＜390 nm），因为它的带隙为 3.2 eV[6-8]。解决上述固有限制的方法有表面改性[9]、离子掺杂[2]、贵金属沉积[10-11]和设计异质结[12]等方法。在这些方法中，异质原子掺杂 TiO_2 因能明显提高光催化活性而受到广泛关注。各种金属或非金属离子如 N、C、Fe、Mo、V、Co、Cu 等都被用来掺杂 TiO_2，从而提高其光催化活性。与其他离子相比，Fe 离子掺杂 TiO_2 更有利于提高 TiO_2 的光催化性能，因为 Fe^{3+} 离子半径（0.65 Å）与 Ti^{4+} 离子半径（0.68 Å）相似，所以很容易将 Fe^{3+} 掺杂到 TiO_2 晶格中[13-15]并促进电荷的转移[16-17]。此外，铁是地壳中最丰富的元素之一，廉价无毒。由于铁离子的上述优点，许多研究都集中于使用不同的方法制备铁掺杂的 TiO_2（如辐射诱导法[18]、溶胶凝胶法[19]）以获得具有提高光催化活性的材料。然而，这些 Fe 掺杂策略通常需要精确控制合成条件或多步操作[20]。此外，Fe 掺杂 TiO_2 的性能与材料的合成方法、铁前驱体、Fe 掺杂浓度、孔隙率、粒度和形貌都有很大的关系。因此，迫切需要设计一种简单有效的高性能掺铁二氧化钛的方法。

金属有机骨架材料（MOFs）因其比表面积大、化学可裁剪性好、骨架密度低、孔径大、孔隙可调等优点而受到广泛关注[21-25]。近年来，MOFs 作为模板在超级电容器和锂离子电池电极的合成、储氢材料和气敏材料等领域得到了迅速发展[26-31]。MOFs 作为牺牲模板或前驱体用于派生各种结构良好、表面积大的多孔材料,但大部分工作都集中在 MOFs 及其复合材料的热分解上。例如,Liu 等[32]首次利用 MOFs 作为模板制备了电化学性能优异的纳米孔碳。以铁基 MOF 模板为原料制备了一种纺锤状多孔的 Fe_2O_3 阳极材料,其特殊的结构使锂离子存储性能得到了极大的提高[33]。使用类似的方法,Koo 等[34]使用 Pd@ZIF-8 作为模板形成了 PdO@ZnO 多相催化剂。Dekrafft 等[35]报道了直接煅烧 MIL-101@TiO_2 制备 Fe_2O_3@TiO_2,并研究了其光催化产氢的性能。因此,使用 MOF 作为模板可以为制备大比表面积的光催化剂提供一种路径。

在本研究中,与传统的牺牲模板煅烧法不同,首次以 MIL-101(Fe)为铁源和模板,采用一步水热法制备了具有八面体形态的 Fe 掺杂 TiO_2(Fe-TiO_2)。Fe^{3+} 在 TiO_2 晶格中高度分散,使得 TiO_2 吸收边明显向可见光区红移。Fe-TiO_2 八面体具有较大的 BET 比表面积(275 m^2/g)、Fe 在 TiO_2 晶格中分散均匀、带隙较小(2.75 eV)等特点,使其具有优异的光催化还原 CO_2 性能。在可见光照射 12 h 时,Fe-TiO_2 光催化还原 CO_2 为 CH_4 的产率约为 2.76 $\mu mol/g$。Fe-TiO_2 样品 500 ℃ 热处理后,得到的 Fe-TiO_2-500 也具有高的 BET 比表面积(202 m^2/g),平均孔径为 3.9 nm,带隙约为 2.42 eV,使得更多的可见光被吸收,从而进一步提高了其光催化活性。在 Fe-TiO_2-500 上,负载 Pt 作为助催化剂进一步提高了 CH_4 的产率,在可见光照射下反应 12 h,CH_4 产率约为 7.73 $\mu mol/g$。

5.2 实验部分

5.2.1 MIL-101(Fe)的合成

取 $FeCl_3 \cdot 6H_2O$(0.5 g)和对苯二甲酸(0.15 g)溶于 DMF(12 mL)中,所得悬浮液经超声分散,移至聚四氟乙烯高压釜内,130 ℃ 加热 24 h,反应完成后在 10 000 r/min 离心 2 min,并用 N,N-二甲基甲酰胺(DMF)和乙醇分别洗涤 3 次。最后,将合成的 MIL-101(Fe)在 60 ℃ 的烘箱中干燥一夜。

5.2.2 Fe-TiO_2,Fe-TiO_2-500 和 TiO_2 的合成

将 MIL-101(Fe)(5.0 mg)溶于 10 mL 乙醇中,加入钛酸四丁酯(50 μL)。磁搅拌 20 min 使混合物均匀分散,再将 700 μL 去离子水和 20 μL HF 混合并搅拌

10 min,最后转入 25 mL 聚四氟乙烯内衬不锈钢高压釜中,180 ℃加热 20 h。将样品离心分离,用去离子水和无水乙醇反复洗涤,60 ℃真空干燥 24 h,所得粉末命名为 Fe-TiO$_2$。为了在 CO$_2$ 还原开始前去除残留的 MIL-101 和有机溶剂,将 Fe-TiO$_2$ 在 500 ℃的空气中煅烧 6 h,得到的产物为 Fe-TiO$_2$-500。纯 TiO$_2$ 纳米颗粒的合成与 Fe-TiO$_2$ 的合成非常相似,只是去除了 MIL-101(Fe)(5.0 mg)。

5.3 结果与讨论

图 5-1 阐述了八面体 Fe-TiO$_2$ 和 Fe-TiO$_2$-500 的合成过程。MIL-101(Fe)八面体由溶剂热法制备。MIL-101(Fe)八面体、HF 和钛酸四丁酯再次溶剂热形成 Fe-TiO$_2$。为了去除残余的 MIL-101(Fe),将 Fe-TiO$_2$ 在空气中以 500 ℃烧结 6 h 获得 Fe-TiO$_2$-500 样品。

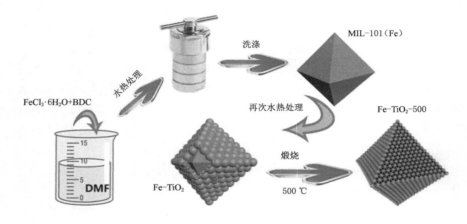

图 5-1　Fe-TiO$_2$ 和 Fe-TiO$_2$-500 八面体的合成过程示意图

图 5-2 为制备的 MIL-101(Fe)、TiO$_2$、Fe-TiO$_2$ 和 Fe-TiO$_2$-500 的 XRD 图谱。从中可以看出,MIL-101(Fe)前驱体的衍射峰值与先前报告的结果完全一致[36]。在 180 ℃下用钛酸四丁酯和 HF 水溶液处理 20 h 后,MIL-101(Fe)的特征峰值消失,在 25.28°、37.92°、48.06°、53.98°、55.1°和 62.74°位置出现的衍射峰与锐钛矿相 TiO$_2$(JCPDS 21-1272)的(101)、(004)、(200)、(105)、(211)和(204)晶面一一对应。这与未使用 MIL-101(Fe)制备的纯 TiO$_2$ 纳米颗粒相似。此外,相比较于纯 TiO$_2$,Fe-TiO$_2$ 的衍射峰显著拓宽,表明 Fe-TiO$_2$ 的尺寸更小。Fe-TiO$_2$ 的 XRD 图谱中 MIL-101(Fe)的特征峰消失,

说明 MIL-101(Fe)完全转化为 Fe-TiO$_2$。热处理后,Fe-TiO$_2$-500 和 Fe-TiO$_2$ 的 XRD 图谱相似,只是强度略有增加。未观察到与铁相关的特征峰。热处理后,在 24°～27°范围内(101)衍射峰对应的 2θ 衍射角有稍微向右偏移[图 5-2(b)],说明 Fe-TiO$_2$ 中有更多的 Ti^{4+} 被 Fe^{3+} 取代。图 5-2(c)所示为这些样品的拉曼位移。TiO$_2$、Fe-TiO$_2$ 和 Fe-TiO$_2$-500 的拉曼光谱很相似,在低频区域有 144.8 cm^{-1}、396.3 cm^{-1}、513.2 cm^{-1} 和 636.7 cm^{-1} 这四个峰值,这些是锐钛矿 TiO$_2$ 的特征峰。在 Fe-TiO$_2$ 和 Fe-TiO$_2$-500 样品中未观察到与 MIL-101(Fe)前体或碳对应的特征峰值,从而确认 Fe-TiO$_2$ 和 Fe-TiO$_2$-500 八面体颗粒中不存在碳或 MIL-101(Fe)残留。

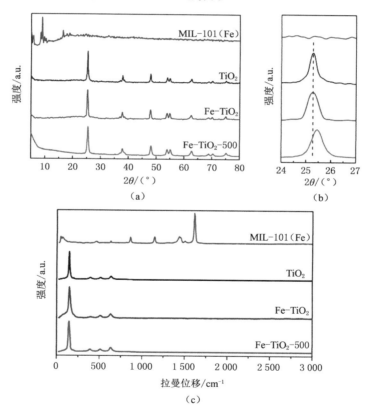

图 5-2 样品的 XRD 图谱、XRD 局部图和拉曼光谱

图 5-3 为制备的 MIL-101 和 Fe-TiO$_2$ 样品的形貌图。所获得的 MIL-101(Fe)具有均匀的八面体形貌,八面体表面光滑,平均粒径为 300～500 nm

[图 5-3(a)]。MIL-101 水热处理后,Fe-TiO$_2$ 样品的 SEM 图像[图 5-3(b)]表明,样品保留了 MIL-101 前驱体的八面体形貌,八面体颗粒大小增加到 1~2 μm,表面变得更粗糙[图 5-3(c)],这有利于随后的光催化反应提供更多的反应位点,提高其光催化性能。单个 Fe-TiO$_2$ 八面体[图 5-3(c)]的插图的高倍 SEM 图像表明,每个八面体都由纳米粒子组成,这为 TiO$_2$ 光催化剂提供了一个较大的表面积。在空气中 500 ℃ 退火后,八面体的形态和大小没有明显变化[图 5-3(d)],表明 Fe-TiO$_2$ 具有很好的稳定性。

图 5-3　样品的 SEM 图

利用透射电子显微镜(TEM)对制备的 Fe-TiO$_2$ 进行进一步微观结构和形貌分析,结果如图 5-4 所示。通过 TEM 观察八面体结构,Fe-TiO$_2$ 样品呈菱形或立方状[图 5-4(a)、(b)]。仔细观察 Fe-TiO$_2$ 八面体,可以看到深色边缘与明亮中心之间有明显的对比,这说明 Fe-TiO$_2$ 八面体具有中空结构。图 5-4(b)的插图是八面体 Fe-TiO$_2$ 的相应选区电子衍射(SAED)图,这表明单个八面体为多晶结构。更高分辨率的 TEM 图[图 5-4(c)]显示,每个八面体 Fe-TiO$_2$ 由许多尺寸为 10～20 nm 的小纳米粒子组成。如图 5-4(c)中插图所示,高分辨率透射电子显微镜(HRTEM)图清楚地显示出清晰的晶格条纹,晶面间距约为 0.356 nm,对应于 TiO$_2$ 锐钛矿相的(101)面。通过能量色散 X 射线谱(EDS)进行元素面扫,Fe-TiO$_2$ 样品中 Ti、O 和 Fe 分布均匀[图 5-4(e)～(g)]。Fe、Ti 和 O 的含量分别为 6.8%、32.0%和 61.2%。Ti 和 O 元素的原子比为 1∶9,接近 TiO$_2$ 的理论原子比。Fe 元素来自 MIL-101(Fe)前体。值得注意的是,在水热反应中加入适当的 MIL-101(Fe)时,溶液中没有 TiO$_2$ 沉淀。所有 TiO$_2$ 纳米粒子在 MIL-101 表面生长,同时 MIL-101 晶体被 HF 腐蚀,导致 MIL-101 消失,并形成中空的 TiO$_2$ 八面体。

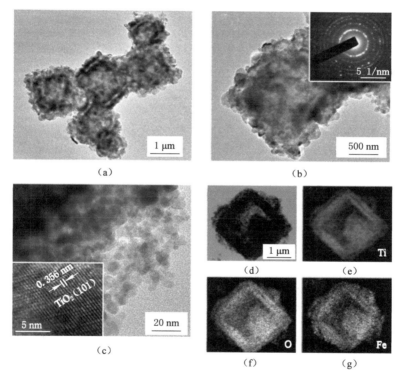

图 5-4　样品的 TEM 图及单个八面体 Fe-TiO$_2$ 的 EDS 元素面扫图

对这三个样品进行热重分析,结果如图 5-5(a)所示。从图中可以看,MIL-101(Fe)有三个明显的失重过程,分别为 50～85 ℃范围内表面吸附水分子的去除,DMF 溶剂在 100 ℃以下的去吸附,MIL-101(Fe)在 330～450 ℃之间的分解,500 ℃以上的轻微重量损失是由氧化铁的生成引起的。加热到800 ℃,MIL-101(Fe)的总质量损失高达76%。对于纯 TiO_2 纳米粒子,在800 ℃时的总失重率仅为9%,这与纳米粒子表面的乙醇和水解吸附有关。Fe-TiO_2 八面体的TG曲线与 TiO_2 相似,失重量略微增加(≤14%),这是由MIL-101前驱体微量残留引起的。图5-5(b)所示为MIL-101(Fe)、Fe-TiO_2 八面体和 TiO_2 的FT-IR光谱。TiO_2 在403.12 cm^{-1} 的峰是由Ti—O—Ti拉伸振动引起的[37]。MIL-101(Fe)有两个主峰1 387.37 cm^{-1} 和1 603.86 cm^{-1},这些主峰由MIL-101(Fe)中 COO^- 的不对称和对称振动引起,与文献中报道的峰值一致[38]。八面体Fe-TiO_2 的FT-IR光谱与 TiO_2 相似,并且没有MIL-101(Fe)的峰值,这表明MIL-101(Fe)没有残留。此外,在1 638.65 cm^{-1} 和3 451.80 cm^{-1} 出现的峰,是由于吸附水分子的弯曲和拉伸振动造成的[39]。TG和FT-IR结果表明,在已生成的Fe-TiO_2 中没有MIL-101(Fe)前驱体残留。

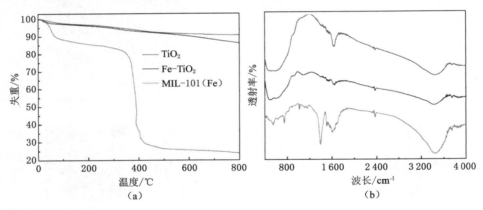

图5-5 样品的TG曲线和FT-IR图谱

图5-6所示为Fe-TiO_2 和Fe-TiO_2-500的 N_2 吸附/脱附等温曲线和孔径分布(插图)。根据BDDT的分类,两个样品表现出H3型迟滞回线,表明存在介孔结构。根据BET方法,Fe-TiO_2 和Fe-TiO_2-500的比表面积分别是275 m^2/g 和202 m^2/g,远远高于纯 TiO_2 的比表面积(43 m^2/g)。使用BJH方法计算其孔径分布,得到Fe-TiO_2 和Fe-TiO_2-500的孔径大小分别约为3.7 nm和3.9 nm。热处理时候,吸附的有机溶剂的挥发和煅烧后残留的MIL-101(Fe)的分解导致Fe-TiO_2-500孔径增大。Fe-TiO_2-500的BET表面积减少是由于高温下 TiO_2 纳米粒子的晶体生长和孔径的增加导致的。

第 5 章 Fe^{3+} 掺杂 TiO_2 八面体的制备及其光催化还原 CO_2 研究

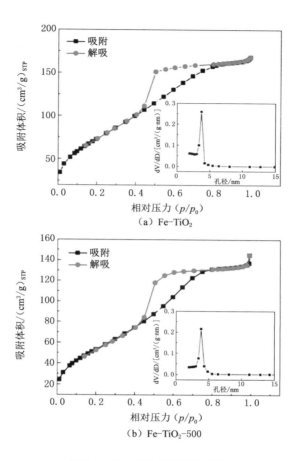

图 5-6 N_2 吸附-脱附等温曲线

X 射线光电子能谱（XPS）分析表明，在合成的 $Fe-TiO_2$ 八面体中存在 Ti、O 和 Fe[图 5-7(a)]。在 458.5 eV 和 464.2 eV 附近出现的峰[图 5-7(b)]，对应 TiO_2 的 Ti $2p_{3/2}$ 和 Ti $2p_{1/2}$ 特征峰。Fe 在 711.3 eV 和 725.0 eV 处的两个主峰分别属于 Fe^{3+} 的 Fe $2p_{3/2}$ 和 Fe $2p_{1/2}$[图 5-7(c)]。与 Fe_2O_3（Fe $2p_{3/2}$ 为 710.7 eV）相比，Fe 2p 在 $Fe-TiO_2$ 中有略微向更高结合能的方向移动，说明一些 Fe^{3+} 取代了 TiO_2 的 Ti^{4+}，形成了 Fe—O—Ti 键[40]。此外，在 719.6 eV 和 733.6 eV 处有两个峰，这是 Fe^{3+} 的卫星峰[41]。O 1s XPS 光谱可以分解为两个峰[图 5-7(d)]，一个峰位于 529.8 eV，可以归为 TiO_2 的晶格氧[42]。O 1s 的另一个峰在 531.4 eV 处，可能属于表面基团中的化学吸附氧（即 OH、

H_2O)[43]。经计算,Fe-TiO_2 八面体表面 Fe、Ti 和 O 含量分别为 7.0%、30.9% 和 62.1%,这与 EDS 结果非常相似。

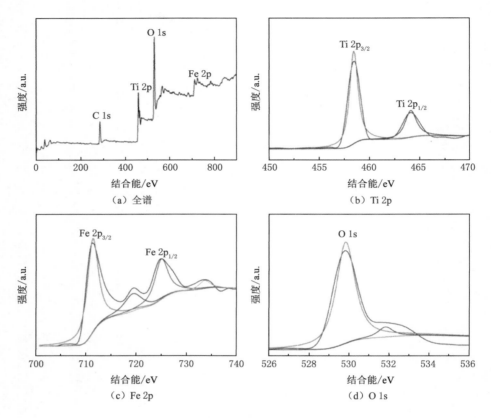

图 5-7　Fe-TiO_2 八面体的 XPS 谱图

图 5-8 为样品的外-可见吸收(UV-vis)光谱图。MIL-101(Fe)的吸收边位于 600 nm 左右。与纯锐钛矿相 TiO_2 相比,Fe-TiO_2 八面体在 400～600 nm 波长处的吸光度大大提高。热处理后吸收边继续红移。由切线截距计算得出,纯 TiO_2、MIL-101(Fe)、Fe-TiO_2 和 Fe-TiO_2-500 的光学带隙分别为 3.30 eV、2.80 eV、2.75 eV 和 2.42 eV[图 5-8(b)]。当将 Fe 掺杂到 TiO_2 中时,TiO_2 的带隙减小。纯 TiO_2 的颜色为白色,而 Fe-TiO_2 的颜色为黄色。带隙越小,表明对可见光的响应越强,因此有望获得具有出色光催化活性的 Fe-TiO_2 八面体。

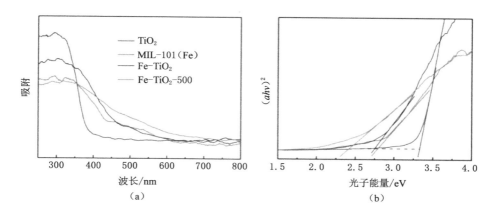

图 5-8　样品的 UV-vis 光谱图和 Kubelka-Munk 函数与光子能量的对应关系

利用光还原[44]、电化学还原[45]或甲烷化[46]将 CO_2 转化为有价值的燃料，为缓解全球能源危机和环境问题提供了一种有希望的途径。本书将制备的 Fe-TiO_2 八面体用于水为还原剂的光催化 CO_2 还原体系。在可见光 ($\lambda > 420$ nm)照射下，在含有水和 CO_2 的气-固系统中进行光催化还原实验，结果如图 5-9 所示。图 5-9(a)显示，连续光照 12 h 后，Fe-TiO_2 八面体催化剂还原 CO_2 生成 CH_4 的速率为 2.76 μmol/g。相比之下，Fe-TiO_2-500 还原 CO_2 表现出更高的光催化活性，CH_4 的生成速率为 5.68 μmol/g。通过负载 Pt(1%)作为助催化剂，Fe-TiO_2-500 八面体还原 CO_2 生成 CH_4 的速率提高至 7.73 μmol/g，这是因为引入 Pt 后，Pt 作为电子陷阱，有利于光生电子-空穴对的分离。在可见光下，Fe-TiO_2、Fe-TiO_2-500 和 Fe-TiO_2-500/Pt 还原 CO_2 生成 CH_4 的速率分别约为 0.23 μmol/g、0.47 μmol/g 和 0.65 μmol/g。Fe-TiO_2-500/Pt 的 CH_4 产率远高于其他掺杂 Fe 的 TiO_2 光催化材料[47]，这可以归因于 TiO_2 和 Pt 之间高的电荷转移效率、高的比表面积以及独特的中空结构。这三种光催化剂还原 CO_2、CH_4 与 O_2 的产率比接近理论值 ($CO_2 + 2H_2O \longrightarrow CH_4 + 2O_2$ 中 CH_4 与 O_2 摩尔比 1∶2)，这表明没有其他还原产物如 CO 和 CH_2O 等产生。CO_2 光还原实验在黑暗条件下或在没有光催化剂的情况下均未形成 CH_4，证明了在光催化剂作用下 CO_2 的还原反应是由光驱动的。电子顺磁共振(EPR)可以提供有关顺磁性掺杂离子所在晶格位点的灵敏光谱信息。Fe-TiO_2 和 Fe-TiO_2-500 八面体的 EPR 光谱如图 5-9(c)所示。两个样品的最强峰出现在 g=1.99 处，这是由于 TiO_2 晶格中 Fe^{3+} 取代了

Ti^{4+}[48],该结果证实 Fe^{3+} 被成功地掺杂到 TiO_2 晶格中。上述 EPR 信号经过 500 ℃ 热处理后明显增强,这是由于热处理后 TiO_2 晶格中加入了更多的 Fe^{3+}。图 5-9(d) 所示为样品的光致发光(PL)光谱,与 $Fe-TiO_2$ 相比,$Fe-TiO_2$-500 的 PL 强度略有下降,说明更多的 Fe^{3+} 掺杂到氧化钛后,促进了光生载流子的分离,降低了载流子的复合效率,这与光催化的实验结果一致。

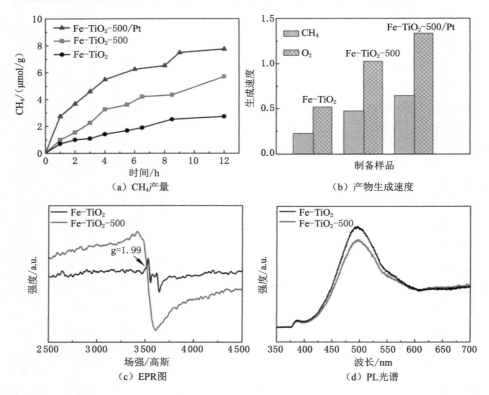

图 5-9 样品还原 CO_2 的 CH_4 产量、产物生成速度、EPR 图和 PL 光谱

图 5-10 给出了 Fe 掺杂对 TiO_2 能带结构改变的示意图。Fe 掺杂对 TiO_2 光催化活性的影响可以解释为:① 掺杂材料的能级(Fe^{2+}/Fe^{3+})低于 TiO_2 的导带能级,这使得导带向下移动,从而使 TiO_2 的带隙变小。② Fe 掺杂还会通过 Fe 3d、O 2p 和 Ti 3d 轨道之间的杂化,从而导致价带发生变化,然后使带隙变窄。③ Fe 掺杂会在 TiO_2 的带隙中引起一个额外的缺陷中心[49],它既可以作为空穴陷阱,也可以作为电子陷阱,从而延长其寿命。因此,$Fe-TiO_2$ 样品可以被可见光激活,并产生更多的光生电子来还原 CO_2。此

外,采用 MIL-101(Fe)作为模板可使产物获得大的比表面积,使得 Fe-TiO$_2$ 样品具有大的比表面积、高孔隙率、孔结构丰富,这些特征能够大大促进 CO$_2$ 的吸附,增加 CO$_2$ 光还原反应的活性位点,因此提高了 Fe-TiO$_2$ 的光催化活性。

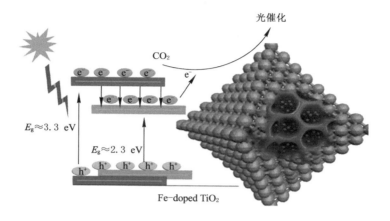

图 5-10 Fe-TiO$_2$ 样品的能带结构和光催化还原 CO$_2$ 示意图

5.4 本章小结

以 MIL-101(Fe)为模板和铁离子源,使用一步水热法成功制备了 Fe^{3+} 掺杂的 TiO$_2$(Fe-TiO$_2$)八面体。TiO$_2$ 颗粒的生长和 MIL-101(Fe)的腐蚀是在氟离子的驱动下同时发生的。所得到的 Fe-TiO$_2$ 在热处理前后 Fe 掺杂均匀,比表面积大(Fe-TiO$_2$:274 m^2/g,Fe-TiO$_2$-500:202 m^2/g)。最后,Fe-TiO$_2$-500 八面体表现出优异的光催化活性,在可见光照射下反应 12 h,载 Pt 后 CH$_4$ 的产率约为 7.73 μmol/g。这项工作为设计和制备具有高光催化活性的新型光催化材料提供了方便。

参考文献

[1] FUJISHIMA A, HONDA K. Electrochemical photolysis of water at a semiconductor electrode[J]. Nature, 1972, 238(5358):37-38.

[2] RAUF M A, MEETANI M A, HISAINDEE S. An overview on the pho-

tocatalytic degradation of azo dyes in the presence of TiO_2 doped with selective transition metals[J].Desalination,2011,276(1/2/3):13-27.

[3] WANG C Y,BÖTTCHER C,BAHNEMANN D W,et al.A comparative study of nanometer sized Fe(Ⅲ)-doped TiO_2 photocatalysts: synthesis, characterization and activity[J].Journal of materials chemistry,2003,13(9):2322-2329.

[4] AMBATI R,GOGATE P R.Ultrasound assisted synthesis of iron doped TiO_2 catalyst[J].Ultrasonics sonochemistry,2018,40:91-100.

[5] ZHOU M H,YU J G,CHENG B.Effects of Fe-doping on the photocatalytic activity of mesoporous TiO_2 powders prepared by an ultrasonic method[J].Journal of hazardous materials,2006,137(3):1838-1847.

[6] ZHAO W,MA W H,CHEN C C,et al.Efficient degradation of toxic organic pollutants with $Ni_2O_3/TiO_{(2-x)}B_x$ under visible irradiation[J]. Journal of the American chemical society,2004,126(15):4782-4783.

[7] XIE Y,ALI G,YOO S H,et al.Sonication-assisted synthesis of CdS quantum-dot-sensitized TiO_2 nanotube arrays with enhanced photoelectrochemical and photocatalytic activity[J].ACS applied materials and interfaces,2010,2(10):2910-2914.

[8] MORADI V,JUN M B G,BLACKBURN A,et al.Significant improvement in visible light photocatalytic activity of Fe doped TiO_2 using an acid treatment process[J].Applied surface science,2018,427:791-799.

[9] LOW J,CHENG B,YU J G.Surface modification and enhanced photocatalytic CO_2 reduction performance of TiO_2: a review[J].Applied surface science,2017,392:658-686.

[10] KUMAR S G,RAO K S R K.Comparison of modification strategies towards enhanced charge carrier separation and photocatalytic degradation activity of metal oxide semiconductors (TiO_2, WO_3 and ZnO)[J].Applied surface science,2017,391:124-148.

[11] LU Q P,ZHU L J,HAN S M,et al.Photocatalytic synthesis of gold nanoparticles using TiO_2 nanorods: a mechanistic investigation[J]. Physical chemistry chemical physics:PCCP,2019,21(34):18753-18757.

[12] HE F,MENG A Y,CHENG B,et al.Enhanced photocatalytic H_2-production activity of WO_3/TiO_2 step-scheme heterojunction by graphene

modification[J].Chinese journal of catalysis,2020,41(1):9-20.

[13] HARIFI T,MONTAZER M.Fe^{3+}:Ag/TiO_2 nanocomposite:Synthesis, characterization and photocatalytic activity under UV and visible light irradiation[J].Applied catalysis A:general,2014,473:104-115.

[14] MAKHATOVA A,ULYKBANOVA G,SADYK S,et al.Degradation and mineralization of 4-tert-butylphenol in water using Fe-doped TiO_2 catalysts[J].Scientific reports,2019,9:19284.

[15] FARHANGI N,CHOWDHURY R R,MEDINA-GONZALEZ Y,et al. Visible light active Fe doped TiO_2 nanowires grown on graphene using supercritical CO_2 [J]. Applied catalysis B:environmental, 2011, 110:25-32.

[16] KIM T H, RODRÍGUEZ-GONZÁLEZ V, GYAWALI G, et al. Synthesis of solar light responsive Fe,N co-doped TiO_2 photocatalyst by sonochemical method[J].Catalysis today,2013,212:75-80.

[17] PANG Y L,ABDULLAH A Z.Fe^{3+} doped TiO_2 nanotubes for combined adsorption-sonocatalytic degradation of real textile wastewater[J].Applied catalysis B:environmental,2013,129:473-481.

[18] BZDON S,GÓRALSKI J,MANIUKIEWICZ W,et al.Radiation-induced synthesis of Fe-doped TiO_2:characterization and catalytic properties[J].Radiation physics and chemistry,2012,81(3):322-330.

[19] AN W J,WANG W N,RAMALINGAM B,et al.Enhanced water photolysis with Pt metal nanoparticles on single crystal TiO_2 surfaces[J]. Langmuir:the ACS journal of surfaces and colloids, 2012, 28 (19):7528-7534.

[20] VALERO-ROMERO M J,SANTACLARA J G,OAR-ARTETA L,et al.Photocatalytic properties of TiO_2 and Fe-doped TiO_2 prepared by metal organic framework-mediated synthesis[J].Chemical engineering journal,2019,360:75-88.

[21] LONG J R,YAGHI O M.The pervasive chemistry of metal-organic frameworks[J].Chemical society reviews,2009,38(5):1213-1214.

[22] LIU Q, ZHOU B B, XU M, et al. Integration of nanosized ZIF-8 particles onto mesoporous TiO_2 nanobeads for enhanced photocatalytic activity[J].RSC advances,2017,7(13):8004-8010.

[23] HONG D Y, HWANG Y K, SERRE C, et al. Porous chromium terephthalate MIL-101 with coordinatively unsaturated sites: surface functionalization, encapsulation, sorption and catalysis[J]. Advanced functional materials, 2009, 19(10): 1537-1552.

[24] FENG Y, CHEN Q, CAO M J, et al. Defect-tailoring and titanium substitution in metal-organic framework UiO-66-NH$_2$ for the photocatalytic degradation of Cr(Ⅵ) to Cr(Ⅲ)[J]. ACS applied nano materials, 2019, 2(9): 5973-5980.

[25] LI S Y, MENG S, ZOU X Q, et al. Rhenium-functionalized covalent organic framework photocatalyst for efficient CO$_2$ reduction under visible light[J]. Microporous and mesoporous materials, 2019, 285: 195-201.

[26] YANG S J, KIM T, IM J H, et al. MOF-derived hierarchically porous carbon with exceptional porosity and hydrogen storage capacity[J]. Chemistry of materials, 2012, 24(3): 464-470.

[27] CHAIKITTISILP W, HU M, WANG H J, et al. Nanoporous carbons through direct carbonization of a zeolitic imidazolate framework for supercapacitor electrodes[J]. Chemical communications (Cambridge, England), 2012, 48(58): 7259-7261.

[28] HUANG G, ZHANG F F, DU X C, et al. Core-shell NiFe$_2$O$_4$@TiO$_2$ nanorods: an anode material with enhanced electrochemical performance for lithium-ion batteries[J]. Chemistry: a european journal, 2014, 20(35): 11214-11219.

[29] LÜ Y Y, ZHAN W W, HE Y, et al. MOF-templated synthesis of porous Co$_3$O$_4$ concave nanocubes with high specific surface area and their gas sensing properties[J]. ACS applied materials and interfaces, 2014, 6(6): 4186-4195.

[30] FENG Y, CHEN Q, JIANG M Q, et al. Tailoring the properties of UiO-66 through defect engineering: a review[J]. Industrial and engineering chemistry research, 2019, 58(38): 17646-17659.

[31] LIU Q, LOW Z X, FENG Y, et al. Direct conversion of two-dimensional ZIF-L film to porous ZnO nano-sheet film and its performance as photoanode in dye-sensitized solar cell[J]. Microporous and mesoporous materials, 2014, 194: 1-7.

[32] LIU B, SHIOYAMA H, AKITA T, et al. Metal-organic framework as a template for porous carbon synthesis[J]. Journal of the american chemical society, 2008, 130(16): 5390-5391.

[33] XU X D, CAO R G, JEONG S, et al. Spindle-like mesoporous α-Fe_2O_3 anode material prepared from MOF template for high-rate lithium batteries[J]. Nano letters, 2012, 12(9): 4988-4991.

[34] KOO W T, JANG J S, CHOI S J, et al. Metal-organic framework templated catalysts: dual sensitization of PdO-ZnO composite on hollow SnO_2 nanotubes for selective acetone sensors[J]. ACS applied materials and interfaces, 2017, 9(21): 18069-18077.

[35] DEKRAFFT K E, WANG C, LIN W B. Metal-organic framework templated synthesis of Fe_2O_3/TiO_2 nanocomposite for hydrogen production[J]. Advanced materials, 2012, 24(15): 2014-2018.

[36] SKOBELEV I Y, SOROKIN A B, KOVALENKO K A, et al. Solvent-free allylic oxidation of alkenes with O_2 mediated by Fe- and Cr-MIL-101[J]. Journal of catalysis, 2013, 298: 61-69.

[37] NAGARAJU G, RAVISHANKAR T N, MANJUNATHA K, et al. Ionothermal synthesis of TiO_2 nanoparticles: photocatalytic hydrogen generation[J]. Materials letters, 2013, 109: 27-30.

[38] ISLAM D A, CHAKRABORTY A, ACHARYA H. Fluorescent silver nanoclusters(AgNCs) in the metal-organic framework MIL-101(Fe) for the catalytic hydrogenation of 4-nitroaniline[J]. New journal of chemistry, 2016, 40(8): 6745-6751.

[39] DING Z, LU G Q, GREENFIELD P F. Role of the crystallite phase of TiO_2 in heterogeneous photocatalysis for phenol oxidation in water[J]. The journal of physical chemistry B, 2000, 104(19): 4815-4820.

[40] XU Z, FAN Z W, SHI Z, et al. Interface manipulation to improve plasmon-coupled photoelectrochemical water splitting on α-Fe_2O_3 photoanodes[J]. ChemSusChem, 2018, 11(1): 237-244.

[41] YAMASHITA T, HAYES P. Analysis of XPS spectra of Fe^{2+} and Fe^{3+} ions in oxide materials[J]. Applied surface science, 2008, 254(8): 2441-2449.

[42] YANG S X, ZHU W P, JIANG Z P, et al. The surface properties and the

[43] T CAMPBELL C.An XPS study of molecularly chemisorbed oxygen on Ag(111)[J].Surface science,1986,173(2/3):641-646.

[44] KANDY M M.Carbon-based photocatalysts for enhanced photocatalytic reduction of CO_2 to solar fuels[J].Sustainable energy and fuels,2020,4(2):469-484.

[45] YADAV R,AMOLI V,SINGH J,et al.Plasmonic gold deposited on mesoporous $Ti_xSi_{1-x}O_2$ with isolated silica in lattice:an excellent photocatalyst for photocatalytic conversion of CO_2 into methanol under visible light irradiation[J].Journal of CO_2 utilization,2018,27:11-21.

[46] GAC W,ZAWADZKI W,SŁOWIK G,et al.Nickel catalysts supported on silica microspheres for CO_2 methanation[J].Microporous and mesoporous materials,2018,272:79-91.

[47] KAUR N,SHAHI S K,SHAHI J S,et al.Comprehensive review and future perspectives of efficient N-doped,Fe-doped and(N,Fe)-co-doped titania as visible light active photocatalysts[J].Vacuum,2020,178:109429.

[48] TONG T Z,ZHANG J L,TIAN B Z,et al.Preparation of Fe^{3+}-doped TiO_2 catalysts by controlled hydrolysis of titanium alkoxide and study on their photocatalytic activity for methyl orange degradation[J].Journal of hazardous materials,2008,155(3):572-579.

[49] MA J Z,HE H,LIU F D.Effect of Fe on the photocatalytic removal of NO_x over visible light responsive Fe/TiO_2 catalysts[J].Applied catalysis B:environmental,2015,179:21-28.

第6章 Cr$_2$O$_3$/C@TiO$_2$复合材料的制备及其光催化产氢性能研究

6.1 引言

太阳光每秒照射到地球的能量大约是 $1.465×10^{14}$ J,这相当于一座大型发电厂10天的工作产能。虽然太阳能很难完全转化为我们所需要的能源,但是采取有效的方法来提高能源转化效率对人类未来的发展具有重大意义。通过光催化剂的介入将太阳能充分应用于分解水制氢是公认的解决当前问题的有效途径之一,可以缓解日益严重的全球性能源危机,也可以改善全球环境污染所带来的问题[1-3]。自从1972年藤岛和本田发表了关于 TiO$_2$ 电极光解水[4]的论文以来,半导体材料一直备受研究人员的关注。迄今为止,半导体材料(如 CuO[5]、ZnO[6]、SnO$_2$[7]、石墨化氮化碳[8-11]、CdS[12-13]和 BiVO$_4$[14-15])已被用来光催化制氢。与其他半导体光催化材料相比,TiO$_2$ 具有化学稳定性强、成本低、无毒等优点[16-17]。但其光生电子-空穴对复合速率较高,且 TiO$_2$ 带隙较宽($\leqslant$3.2 eV),对可见光吸收能力低,极大地限制了 TiO$_2$ 的在光催化领域的实际应用[18]。

通过离子掺杂[19]、表面修饰[20-21]、负载助催化剂[22-24]以及形成异质结[25-28]等手段能够大幅度提高 TiO$_2$ 的光催化活性。其中,过渡金属阳离子掺杂和过渡金属氧化物复合是增强可见光吸收能力、提高 TiO$_2$ 光催化性能的有效方法之一[29-30]。Cr$_2$O$_3$ 带隙窄,是一种高效的光催化剂材料[31-32]。Irie 等[33]采用浸渍法将 Cr^{3+} 掺杂到 TiO$_2$ 粉末中,改善了其光催化性能。Jun 等[34]合成了 Cr 掺杂的 TiO$_2$ 薄膜,证明其不同的带隙值取决于 TiO$_2$ 薄膜的溅射参数。Wang 等[35]采用溶胶-凝胶法合成了 Cr 和 Co 共掺杂的 TiO_2 粉,研究了其对偶氮的光催化活性。结果表明,Cr 和 Co 共掺杂的 TiO_2 粉对偶氮的光催化活性超过了 TiO_2。上述研究均证实了 Cr 掺杂和 Cr_2O_3 改性可以提高 TiO_2 的光催化活性。

金属有机骨架(MOFs)由于其独特的结构、高的表面积和大量的微孔结构,在许多研究领域受到了相当多的关注[36-37]。使用 MOFs 作为前驱体和模板有望获得结构可调和比表面积大的 MOFs 多孔衍生材料[38-39]。MOFs 的高表面积、大孔隙率和可定制性可以被其衍生物继承,因此,MOFs 衍生材料具有巨大的催化应用潜力。迄今为止,MOFs 衍生材料大多数研究都集中在多孔碳[40-41]、金属氧化物或碳复合金属氧化物材料的制备上[42-44]。通过在不同条件下煅烧 MOFs,以制备的 MOF 衍生物为核、TiO_2 为壳设计新型光催化剂,可以提高反应物的扩散效率,为催化反应提供更多的反应活性位点,从而能够提高 TiO_2 的光催化效率。

本章以稳定的 MOFs 材料 MIL-101(Cr)为前驱物,采用两步法成功制备了 $Cr_2O_3/C@TiO_2$ 复合材料。其中,MIL-101 作为纳米复合材料的模板,使得制备的材料具有 MIL-101 的八面体结构,同时 MIL-101 作为复合材料的 C 源和 Cr 源。$Cr_2O_3/C@TiO_2$ 复合材料的合成过程如图 6-1 所示。首先,采用对苯二甲酸(BDC)作为有机配体和 $Cr(NO)_3$,在溶剂热条件下合成了 MIL-101 八面体;其次以 MIL-101 纳米粒子为核,使用 TiF_4 为前驱体,在水热条件下获得 MIL-101@TiO_2(缩写为 MT)核壳结构;最后将制备的 MT 核壳结构在 500 ℃ N_2 气氛下碳化 5 h,得到核壳型 $Cr_2O_3/C@TiO_2$ 复合纳米材料。

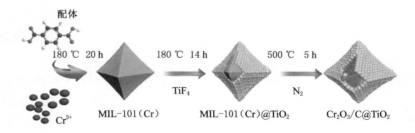

图 6-1 $Cr_2O_3/C@TiO_2$ 的设计合成示意图

6.2 实验部分

6.2.1 八面体 MIL-101(Cr) 的合成

将 0.8 g Cr(NO)$_3$·9H$_2$O 和 0.3 g BDC 的药品粉末溶入 14 mL 的去离子水中。用磁力搅拌器搅拌 15 min,再加入 0.1 mL 氢氟酸(HF,40%),继续搅拌 10 min 以使溶液中的氢氟酸可以均匀分布。用 25 mL 的高压反应釜装入混合后的溶液,进行 180 ℃加热处理,时间为 20 h。待其冷却后离心收集样品,用 DMF 和乙醇分别洗涤 4 次。最后,把所得的样品放在干燥箱中 60 ℃干燥,以备进一步使用。

6.2.2 MIL-101(Cr)@TiO$_2$ 复合材料的合成

将 52 mg MIL-101(Cr)(简称 MT)粉末通过磁力搅拌分散在 10 mL 无水乙醇溶剂中,在搅拌中使用移液枪加入 3 mL 0.022 mol/L 的四氟化钛水溶液。继续搅拌 5 min 后,用 25 mL 的聚四氟乙烯内衬不锈钢高压反应釜装入混合后的溶液,进行 180 ℃加热处理,时间为 14 h。最后将得到的沉淀物用去离子水和乙醇溶剂依次洗涤 4 次,置于 60 ℃烘箱中烘干。

6.2.3 Cr$_2$O$_3$/C@TiO$_2$ 复合材料的制备

将 0.1 g MT 粉末放入 4 个相同规格的石英坩埚中,分别在 400 ℃、500 ℃、600 ℃、700 ℃下进行 N$_2$ 保护加热处理(分别命名为 MT400、MT500、MT600、MT700)。加热和冷却速率控制在 2 ℃/min,保温时间为 5 h。

6.3 结果与讨论

采用扫描电子显微镜(SEM)对制备的材料进行表面形貌分析。图 6-2(a)是通过一步水热法合成的 MIL-101(Cr) 前驱体的 SEM 照片。从照片中可以看出,制备的 MIL-101(Cr) 颗粒均匀分散,呈现出完整的八面体结构,大小在 500~800 nm。图 6-2(b)和(c)是通过二次水热法后制备的 MT 复合材料的 SEM 照片。从照片中可以看出,水热以后样品依然呈现出规整的八面体结构,颗粒大小略有增加,约为 550~850 nm。从图 6-2(c)中的一个破损特例可以清晰地看出内部的 MIL-101(Cr) 前驱体依旧完好无损,只是表面覆盖了一层紧密的颗粒结构。图 6-2(d)是 MT 的透射电镜(TEM)照片,进一步证实了 MT 为核壳结构。与前驱体 MIL-101(Cr)相比,核壳结构内部的 MIL-101(Cr) 明显缩小,这可能是因为二次水热过程中加入的氟离子对前驱体有一定的腐蚀作用,而且随着水热温度的增加,腐蚀效果越来越明显。本章实验用控制变量法研究了 TiF$_4$ 的加入量、水热反应温度和反应时间对制备的 MT 复合材料形貌的影响,并优化了其制备工艺。

图 6-2 MIL-101(Cr)前驱体、MT 中间体的 SEM 照片和 MT 中间体的 TEM 照片

图 6-3 是改变溶液中 TiF$_4$ 的添加量,其他反应条件不变获得的产物的 SEM 照片。图 6-3(a)是对照实验,没有添加任何氟化物,180 ℃、14 h 溶剂热处理后 MOFs 结构没有任何明显变化。这可以证明 MIL-101(Cr)在 180 ℃ 的溶剂热条件下是稳定的,不会被分解。图 6-3(b)添加的是 40 μL 的 HF,结果发现原本光滑平整的 MIL-101(Cr)八面体的表面出现了一些坑坑洼洼,这证明了 F 离子在溶剂热过程中对 MIL-101(Cr)的腐蚀效果。加入一定量的 TiF$_4$ 后,如图 6-3(c)所示,MIL-101(Cr)八面体表面变粗糙,有颗粒堆积,这

是因 TiF_4 在溶剂热过程中会分解生成 HF 和 TiO_2,TiO_2 颗粒沉积到八面体表面。从没有生长 TiO_2 的 MIL-101 颗粒表面可以进一步证实,HF 会在 MIL-101(Cr)表面腐蚀出一层密密麻麻的小坑,从而为晶体提供生长位点,使得 TiO_2 优先在八面体表面沉积。继续增加 TiF_4 的量后发现[图 6-3(d)],MIL-101(Cr)八面体结构几乎消失,生成了形状不规则的多面体,这是由于 TiF_4 产生的 HF 过量,完全腐蚀掉了 MIL-101,氧化钛颗粒随机沉积聚集长大。综合图 6-3(b)~(d),发现添加 3 mL 0.022 mol/L TiF_4 的效果最好。

图 6-3 对照实验(不添加 TiF_4、40 μL HF)及
变量实验(6 mL 0.022 mol/L TiF_4、9 mL 0.022 mol/L TiF_4)产物 SEM 照片

图 6-4 是改变溶剂热的反应温度,其他反应条件不变获得的产物的 SEM 照片。由图 6-4(a)~(c)可知,反应温度为 120~160 ℃时,反应釜内的热动力能不能完全满足 TiO_2 的形核需求,从而导致了 TiO_2 的不均匀形核。在 180 ℃时,TiO_2 会均匀形核并完整地包覆在 MIL-101(Cr)的表面。当溶剂热温度达到 200 ℃时,溶剂热温度过高,加速了 MIL-101(Cr)的腐蚀,表面生长的 TiO_2 壳层破裂,难以获得形貌均匀的核壳结构,如图 6-4(d)所示。综上,180 ℃为最

佳实验温度。

图 6-4　不同溶剂热温度制备的产物的 SEM 照片

图 6-5 是 180 ℃时不同反应时间获得的产物的 SEM 照片。溶剂热反应的时间对复合材料的合成也至关重要，综合图 6-5(a)～(c)可知，无论是 2 h、4 h 还是 8 h，都因为时间过短而导致反应不够充分，TiO_2 未能在 MIL-101(Cr) 表面均匀分布。如图 6-5(d)所示，水热反应 24 h 后获得的产物与前面提到的 14 h 的产物的形貌相比变化并不明显。因此可以得出结论，在水热反应过程中，反应时间过短不利于 TiO_2 的形核，反应时间过长虽然对形貌影响不大，但是也会增加没有必要的能量消耗。出于综合因素考虑，确定 14 h 为本实验的最佳反应时间。

通过一系列的优化实验最终得出结果，52 mg MIL-101(Cr) 和 10 mL 的无水乙醇中加入 3 mL 0.022 mol/L 的 TiF_4 在 180 ℃ 下加热 14 h 为最佳反应条件。在最佳反应条件下制备的 MT 和 TiO_2、MIL-101(Cr) 的 XRD 如图 6-6 所示。MIL-101(Cr) 的特征峰与文献报道一致，而 TiO_2 的特征峰与标准 PDF 卡片(JCPDS No.12-1272)的一致，为锐钛矿相。合成的 MT 同时拥有 MIL-101(Cr) 和 TiO$_2$ 的特征峰,结合上面的 SEM 照片可以得知 MT 是以 MIL-101(Cr) 为核、以 TiO$_2$ 为壳的核壳型复合材料。

图 6-5　180 ℃时不同反应时间制备的产物 SEM 照片

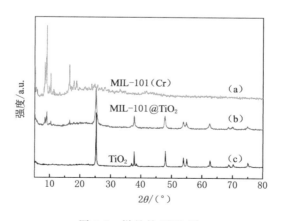

图 6-6　样品的 XRD 图

将制备的 MT 材料在 N_2 保护下,应用不同温度进行加热处理。为了便于说明,400 ℃、500 ℃、600 ℃、700 ℃ 煅烧制备的样品分别标注为 MT400、MT500、MT600 和 MT700。图 6-7 是不同温度热处理制备样品的 XRD 图,与 MT 的 XRD 对比发现,高温碳化后样品在 $2\theta=0°\sim20°$ 处的 MIL-101(Cr) 的特征峰消失了,说明 MT 内部的 MIL-101(Cr) 在高温下发生了转变。MT400 的所有衍射峰与锐钛矿相 TiO_2(JCPDS No.21-1272)吻合,MIL-101(Cr) 的衍射峰消失,这说明 400 ℃ 煅烧获得的样品仅有锐钛矿相的 TiO_2,内核碳化了未形成晶体。当温度增加到 500 ℃ 时,在 $2\theta=33.7°$ 和 $36.4°$ 处出现了两个新的衍射峰,对比标准 PDF 卡片(JCPDS No.38-1479),发现这两个新增的衍射峰对应 Cr_2O_3 的 (104) 和 (110) 面。进一步将碳化温度提高到 600 ℃ 和 700 ℃ 时,Cr_2O_3 对应的 XRD 峰强度和锐度都有增加,这说明样品中 Cr_2O_3 结晶度提高。在 $2\theta=25.3°$ 的特征峰处,因为 Cr^{3+} 的半径比 Ti^{4+} 的半径小,离子半径较小的 Cr^{3+} 的掺入会使得 TiO_2 的晶格结构发生变化,所以随着碳化温度的升高,衍射峰会逐渐向大角度偏移。

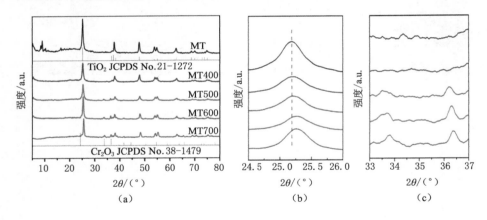

图 6-7 对 MT 进行不同温度下煅烧的 XRD 图和局部放大图

图 6-8 是 MT 在不同温度下煅烧的 SEM 照片,MT400 和 MT500 虽然内部的 MIL-101(Cr) 发生了转变,但是基本形貌相较于 MT 没有变化,依然保持了完整的八面体结构。而 MT600 和 MT700 的结构都有不同程度的坍塌和破碎,结合图 6-7 的 XRD 图谱可知,从 500 ℃ 开始内部的 MIL-101(Cr) 就会转化生成 Cr_2O_3。因此,MT500 不仅获得了 Cr_2O_3 和 TiO_2 的复合材料,而且形成了完整的核-壳结构。

图 6-8　MT400、MT500、MT600 和 MT700 的 SEM 照片

对制备过程中三个阶段的样品分别进行了热重分析,如图 6-9 所示。由图 6-9(a)可以看出,MIL-101(Cr) 在 100 ℃左右 10% 的失重主要是由于样品表面吸附的溶剂分子的挥发以及一些有机物分解所致[45]。300~400 ℃ 之间失重急剧增加,这是由于该温度下 MIL-101(Cr) 的分解所致。在 400℃以上,MIL-101(Cr) 几乎完全被分解,而图 6-9(b)所示 MT 样品在 400 ℃ 过后只损失了 20% 的质量。由于 TiO_2 十分稳定且不可能在 400 ℃ 就发生相变,所以 MT 的失重是因为 MIL-101(Cr) 核的分解。通过此图可以推算出 MT 中 MIL-101(Cr) 和 TiO_2 的比例大概是 1∶4。图 6-9(c)是 MT500 的热重曲线,从曲线上可以看出加热到 800 ℃ 样品的失重约为 88%,这部分的失重主要是由于样品中碳的燃烧。图 6-9(d)所示为 MT500 的拉曼光谱,低频区域出现了 4 个特征峰,分别归属于锐钛矿相 TiO_2 的 $E_{lg(1)}$ (148.3 cm^{-1})、$B_{lg(1)}$ (396.0 cm^{-1})、$A_{1g} + B_{lg(1)}$ (513.8 cm^{-1}) 和 $E_{lg(2)}$ (634.6 cm^{-1}) 模式。另外,1 359 cm^{-1} 和 1 607 cm^{-1} 分别对应碳材料中的 sp^3 缺陷和 sp^2 键合的碳原子[46-47],证明了 MT500 中除了氧化铬和二氧化钛还有碳的存在。所以,拉曼

数据进一步证明了 MT500 是 $Cr_2O_3/C@TiO_2$ 八面体核壳结构复合材料。

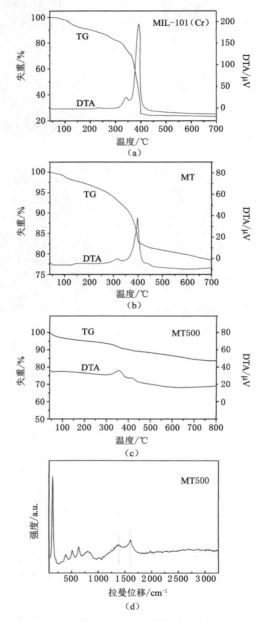

图 6-9　MIL-101(Cr)、MT、MT500 的热重分析及 MT500 的拉曼光谱

第6章 Cr₂O₃/C@TiO₂复合材料的制备及其光催化产氢性能研究

图 6-10 所示为三个样品的红外线光谱,MIL-101(Cr)上出现了大量含氧官能团的特征峰:1 400 cm^{-1} 和 1 610 cm^{-1} 分别对应的是 BDC 配体中羧基不对称的 O—C—O;1 524 cm^{-1} 和 1 700 cm^{-1} 分别代表苯环上的 C=C 和 BDC 配体中羧基对称的 C=O[48]。而在复合过后,这些特征峰会明显减弱。MT500 中这些峰几乎完全消失了,这是因为在 500 ℃ 的 N₂ 保护下 MT 核壳结构内部的 MIL-101(Cr) 碳化形成了 Cr₂O₃/C 的复合材料。

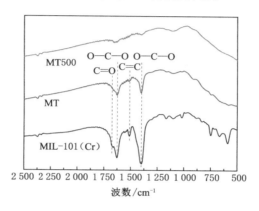

图 6-10 MIL-101(Cr)、MT 和 MT500 的红外线光谱

图 6-11 为制备的 MT500 的 SEM 和 TEM 照片。由图 6-11(a)~(d)可以明显看出,Cr₂O₃/C@TiO₂ 八面体的核壳结构特征,八面体的直径为 500~1 000 nm。壳层的高分辨率 TEM 图像[图 6-11(c)插图]显示,晶格条纹的间距为 0.35 nm,对应于锐钛矿相 TiO₂ 的(101)晶面。图 6-11(d)是单个 MT500 的,空心内核与外壳形成强烈的对比,核为多孔结构,壳层由许多纳米粒子堆积而成。采用用 N₂ 的吸附-脱附曲线进一步研究样品的比表面积,MIL-101(Cr)、MT 和 MT500 的比表面积分别为 2 389 m²/g、454 m²/g 和 65 m²/g,孔径分别为 3.328 nm、3.730 nm 和 3.653 nm。在加热处理的过程中,由于 MIL-101(Cr) 的分解,样品的 BET 比表面积急剧下降。为了进一步表征 MT500 的核壳结构,对该样品做了 EDS 面扫测试,结果如图 6-11(e)~(h) 所示。结果表明,Cr、C、Ti、O 元素在整个八面体结构中不是均匀分布的。因为 C 和 Cr 来自 MIL-101(Cr) 前驱体的碳化,所以 Cr 主要分布在核壳结构的核心部位,Ti 主要分布在 MT500 的壳层上,O 在整个核-壳结构中均匀分布,核心部位的 O²⁻ 会与 Cr³⁺ 结合生成氧化铬。值得注意的是,仍然有少量的 Cr 分布在壳上,这是由于 TiO₂ 晶格中 Cr 元素的掺杂,所得的 Cr-TiO₂ 壳的带

隙将会明显缩小，这有利于增强 $Cr_2O_3/C@TiO_2$ 的光催化活性。

图 6-11　制备的 MT500 的 SEM 和 TEM 照片

MT500 的表面元素组成以及化学价态可以利用 XPS 进行表征。图 6-12(a) 是光催化剂 MT500 的 XPS 的全谱图，可以证实 MT500 结构中存在 Cr、Ti、O、C 元素。MT500 的 Cr 2p XPS 如图 6-12(b)所示，结合能分别为 577.7 eV 和 586.4 eV 的两个峰分别对应于 Cr^{3+} 的 Cr $2p_{3/2}$ 和 Cr $2p_{1/2}$。MT500 的 Ti 2p 的 XPS 如图 6-12(c)所示，在结合能 458.9 eV 和 464.7 eV 处出现两个峰，与文献中记载的 Ti $2p_{3/2}$ 和 Ti $2p_{1/2}$ 的特征信号相一致。

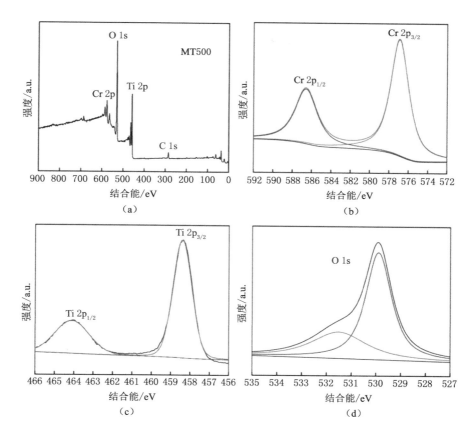

图 6-12 MT500 的 XPS 全谱图及 Cr 2p、Ti 2p 和 O 1s 的 XPS 图

除此之外,对不同温度热处理的样品也进行了 XPS 测试,如图 6-13 所示。可以得知,随着反应温度的逐渐升高,Cr 2p、Ti 2p、O 1s 和 C 1s 的 XPS 都有所偏移。MT500 样品的 Ti 2p 结合能(458.4 eV 和 464.2 eV)低于制备的原始 TiO_2,这可能是由于 Cr^{3+} 取代了 Ti^{4+}。随着煅烧温度的升高,峰值向较小的角度移动,这可能是因为 TiO_2 晶格中有更多的 Ti^{4+} 被 Cr^{3+} 取代。Ti $2p_{3/2}$ 的主峰由 457.7 eV 偏移到 458.4 eV,Ti $2p_{1/2}$ 的主峰由 463.4 eV 偏移到 464.1 eV,主峰向右边移动了约 0.7 eV,表明随着反应温度的升高,TiO_2 晶格中掺入了更多的 Cr^{3+},影响了 TiO_2 的化学状态。MT500 的 C 1s 的 XPS 图谱如图 6-13(d)所示,主峰位于 284.8 eV 处对应的是 C—C 键。其他 288.7 eV 的弱峰为羰基(C=O),对应的是 MIL-101(Cr) 中的对苯二甲酸。随着温

度的升高,这些官能团的峰强度大大减弱,说明 MT 内部的金属有机骨架材料在 N_2 氛围中加热处理会发生碳化。碳化后形成的复合材料不仅保持了原有的核壳型形貌特征,催化性能更会有质的提升。

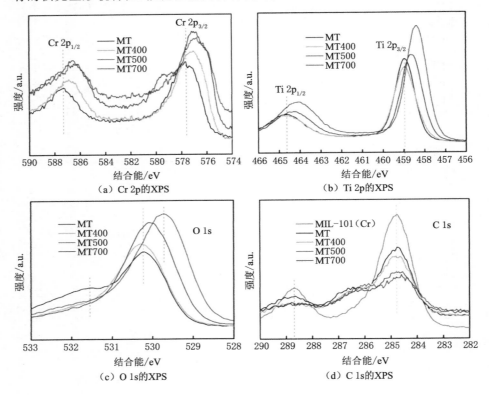

图 6-13 不同温度热处理的样品的 XPS

对催化剂样品的紫外-可见吸收光谱进行了表征,结果如图 6-14 所示。纯 TiO_2 的吸光度边约为 390 nm[图 6-14(a)]。与 TiO_2 相比,MT 和 MT500 的吸收边缘出现了较大的红移。随着 MT 煅烧温度的升高,样品的吸收边缘先向红色可见光区移动,然后向蓝色可见光区轻微移动(图 6-15)。样品的带隙可以用 Kubelka-Munk 函数估计。利用曲线边缘的上集估计原始 TiO_2 和 MT 的带隙约为 3.2 eV 和 3.1 eV,说明与 MIL-101 结合后,TiO_2 的带隙变小了。MT500 有两个明显的吸收边,分别对应于 TiO_2 和 Cr_2O_3。Cr^{3+} 的离子半径为 0.075 5 nm,Ti^{4+} 的离子半径为 0.074 5 nm,由于二者的离子半径非常接近,所以热处理后 Cr^{3+} 很容易进入 TiO_2 晶格中取代 Ti^{4+},并且在 TiO_2 中引入了新的能级,使其带隙减小[49-50]。据估计,MT500 中掺杂 Cr 的 TiO_2 的主带隙约为 2.39 eV,说明其具有良好的可见光吸收能力。

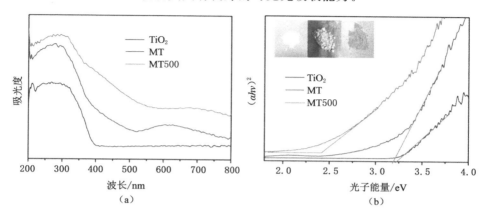

图 6-14　样品的 UV-vis 光谱图和相应的 Kubelka-Munk 函数与光子能量的关系图

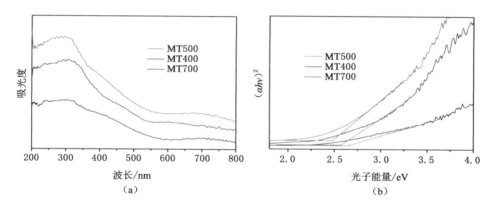

图 6-15　不同热处理温度制备的样品
UV-vis 光谱图和相应的 Kubelka-Munk 函数与光子能量的关系图

此外,MT500、MT400 和 MT700 样品在整个可见光范围内表现出强烈的吸收,这是由于这些纳米复合材料中存在 C。制备样品 TiO_2、MT、MT500 的颜色分别为白色、绿色和黄色。这些结果表明,MT500 样品可能能够吸收更多的光,产生更多的电子-空穴对,因此具有比纯 TiO_2 更好的光催化析氢活性。

以甲醇为牺牲剂,在 300 W 的氙灯照射下对制备的样品进行了光催化制氢的活性检测。在没有光照或没有光催化剂的情况下，分别静置 6 h 后发现没有检测到 H_2 的产生。这个对照实验说明 H_2 是在光照下由光催化剂产生的，从而排除了其他干扰因素。图 6-16(a)是全幅光照下所有实验样品（催化剂）的光催化制氢速率。经过观察发现，MIL-101(Cr) 和 MT 几乎没有光催化活性，说明 MIL-101(Cr) 本身并不是有效的光催化剂。纯 TiO_2 粒子的光催化活性较低，氢气生成速率为 114 $\mu mol/(h\cdot g)$，这是由于光生载流子的快速重组和可见光光谱利用受限所致[51]。在 400 ℃ 碳化后的 MT400 产率明显增加，为 300 $\mu mol/(h\cdot g)$，约为纯二氧化钛的 2.6 倍。随着碳化温度的提高，MT500 的光催化性能达到 446 $\mu mol/(h\cdot g)$，这几乎是纯二氧化钛的 4 倍。然而煅烧温度进一步升高时，MT600、MT700 的光催化活性有降低的趋势。为了研究光催化活性的影响因素，我们又设置了一组对照实验：MT 直接在空气中进行 500 ℃ 热处理（简称 MT-air），结果 MT-air 的析氢速率低至 196 $\mu mol/(h\cdot g)$，只有 MT500 的一半。这些结果表明，在 MT500 的结构中，来自 MIL-101 的 C 的存在有利于增强 MT500 的光催化活性。而 MT500 是在氮气保护下处理的，所以有碳的存在。碳有利于提高催化剂的光催化活性，可以改善电荷的转移，还能促进光生电子-空穴对的分离，这在一些文献中得到了证实[52-54]。

在相同条件下，为了检测 MT500 催化剂的稳定性，对其进行了连续三个周期的析氢检测。如图 6-16(c)所示，经过三次循环后 MT500 的析氢速率仍然维持在 95%，说明 MT500 在光催化 H_2 演化中的稳定性还是很高的，通过 XRD 和 SEM 的表征发现，经过三次（12 h）的性能测试后，材料物相结构和显微形貌并没有发生明显的变化（图 6-17 和图 6-18）。由紫外-可见光谱可以看出，制备的 $Cr_2O_3/C@TiO_2$ 对可见光也有很强的吸收，因此也对催化剂做了可见光下产氢性能的检测。在可见光照射下（$\lambda > 420$ nm）光催化剂的制氢速率如图 6-16(d)所示，Cr_2O_3/C 是通过在 500 ℃ 下 N_2 煅烧原始 MIL-101 得到的，它的 XRD 和 SEM 如图 6-19 所示。Cr_2O_3/C 显示非常低的光催化活性，析氢速率仅为 2 $\mu mol/(h\cdot g)$。MT500 和 MT700 的析氢速率分别是 25.5 $\mu mol/(h\cdot g)$ 和 13.3 $\mu mol/(h\cdot g)$。这说明单一的 Cr_2O_3/C 可见光光催化活性很差，MT500 的催化活性很可能是 TiO_2 与 Cr_2O_3 形成的核壳型复合结构的协同效应。MT700 的光催化活性降低，可能是核-壳纳米结构的坍塌和高温下晶粒的生长造成的。

为了研究光电转换电子-空穴载流子的迁移和分离，对 MT500、MT 和 TiO_2 样品进行了光电流密度、荧光光谱（PL）和电化学阻抗谱（EIS）表征，如图 6-20 所示。光电流测试时间设定为 300 s，初始电压设为 0.4 V vs. Ag/AgCl(1.4 V

第6章 $Cr_2O_3/C@TiO_2$ 复合材料的制备及其光催化产氢性能研究

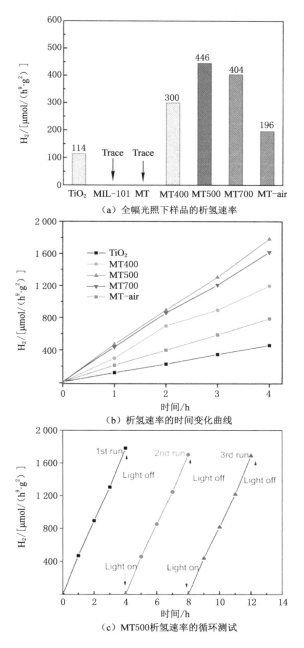

(a) 全幅光照下样品的析氢速率

(b) 析氢速率的时间变化曲线

(c) MT500析氢速率的循环测试

图 6-16 析氢速率及时间变化曲线

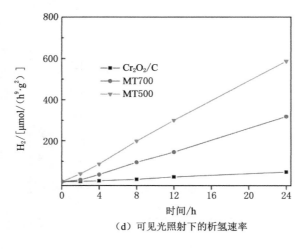

（d）可见光照射下的析氢速率

图 6-16 （续）

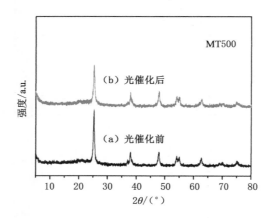

图 6-17 MT500 在光催化制氢前后的 XRD

vs.RHE)，每间隔 25 s 遮挡光源一次，遮光时间为 15 s，光源为 300 W 的氙灯，测试溶液为 1 mmol NaOH 溶液，pH 值为 7。由图 6-20(a) 的曲线发现 MT500 的光电流密度显著高于其他两个样品，这与 MT500 的光催化性能优于 MT 和 TiO_2 相一致。MT500 的光电流密度约为 2.1 mA/cm^2，比原始 TiO_2 和 MT 的光电流密度分别高出 3.5 倍和 5.5 倍。这些结果强烈地表明，MT500 独特的结构促进了光生电子-空穴对的分离。利用电化学阻抗谱 (EIS) 图研究了上述样品的电荷转移电阻，结果如图 6-20(b) 所示。与 MT 和

（a）光催化前　　　　　　　　　　　（b）光催化后

图 6-18　MT500 在光催化制氢前后的 SEM 照片

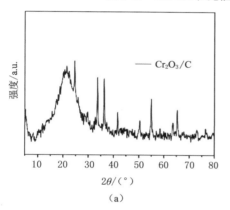 

（a）　　　　　　　　　　　　　　　（b）

图 6-19　Cr_2O_3/C 的 XRD 与 SEM 照片

TiO_2 相比，MT500 样品的半圆更小，MT500 的电荷转移阻抗更低，说明 MT500 的电荷转移阻抗低于其他两种材料。图 6-20(c)是样品的光致发光荧光光谱，与纯 TiO_2 相比，MT500 中的光致发光强度略有下降，说明 MT500 光致发光载流子寿命更长，复合过后的催化剂有效地改善了光生电子-空穴容易复合的问题。

MT500 对光解 H_2 反应的光催化活性高于其他样品，主要有以下三个原因：① 由于 Cr^{3+} 的掺杂，TiO_2 壳层带隙变窄，使合成的 $Cr_2O_3/C@TiO_2$ 具有较宽的可见光响应范围。因此，$Cr_2O_3/C@TiO_2$ 可以获得更多的可见光，从而提高了 MT500 的光解性能。② 来自 MIL-101 的 Cr_2O_3/C 增强了可见光吸收，与 TiO_2 形成 Z-scheme 体系，有利于提高其光催化活性。在 $Cr_2O_3/C@TiO_2$ 中，C 作为固态电子介质。③ Cr_2O_3、C、TiO_2 合成微八面体对析氢反应具有协同作用，为反应提供了更活跃的位点。

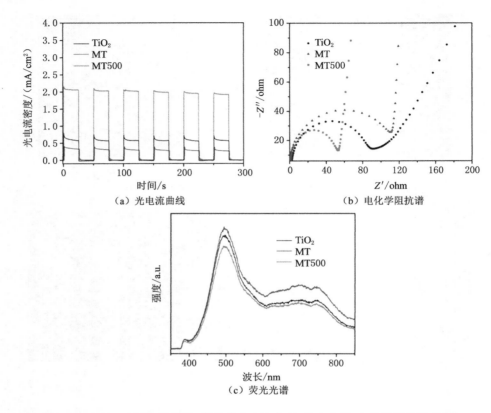

图 6-20 MT500、MT 和 TiO$_2$ 样品的光电流密度、荧光光谱、电化学阻抗谱表征

为了研究 TiO$_2$ 和 Cr$_2$O$_3$ 的能带结构,测试了样品的莫特-肖特基图,如图 6-21 所示。外推的 TiO$_2$ 和 Cr$_2$O$_3$ 平带电势分别为 0.78 V 和 0.45 V(Ag/AgCl,pH=7)。TiO$_2$ 和 Cr$_2$O$_3$ 的平带电势可计算为:

$$E(\text{RHE}) = E(\text{Ag/AgCl}) + E\theta + 0.059\,\text{pH}$$
$$E^\theta(\text{Ag/AgCl}) = 0.197\,\text{V}$$

TiO$_2$ 和 Cr$_2$O$_3$/C 的平带电势分别为 −0.18 V 和 0.16 V(相对于 RHE,pH=0),根据其紫外-可见吸收光谱,计算出原始 TiO$_2$ 的带隙约为 3.2 eV。在 500 ℃ 的空气中煅烧原始 MIL-101 得到纯净的 Cr$_2$O$_3$,其中纯 Cr$_2$O$_3$ 的带隙约为 2.7 eV。掺杂铬的 TiO$_2$ 的价带顶端(VB)随着占据的 Cr^{3+} 含量的提高而升高;而由 Ti 3d 轨道所确定的原导带底部(CB)几乎没有影响[55]。因此,Cr 掺杂 TiO$_2$ 的平带电位仍为 0.18 V。

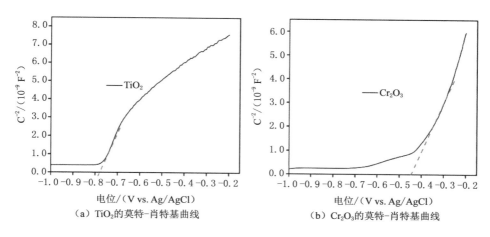

图 6-21　TiO_2 与 Cr_2O_3 样品的莫特-肖特基曲线

利用这些结果，图 6-22 给出了作为 Z 反应体系的 $Cr_2O_3/C@TiO_2$ 光催化活性增强的可能机制示意图。在光照条件下，$Cr-TiO_2$ 和 Cr_2O_3 都会吸收光能激发产生光生载流子。Cr_2O_3 导带上的光电子会优先通过碳链转移到 TiO_2 的价带上，与 TiO_2 中电子转移后留下的空穴重新结合，从而使得 TiO_2 半导体上产生的电子得以释放到催化剂表面的活化位点。此外，TiO_2 的能带宽度由于 Cr^{3+} 的掺杂会变窄，这使得它对光的捕捉也更加敏感。$Cr-TiO_2$ 导带上生成的电子表现出了很强的还原能力，可以将水中氢离子还原为氢气，Cr_2O_3 价带上的空穴也具备一定的氧化能力。这种 Z 字形结构在很大程度上避免了催化剂上光生电子和空穴的复合，十分有利于催化活性的提高。

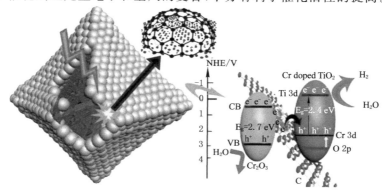

图 6-22　$Cr_2O_3/C@TiO_2$ 复合材料光催化分解水的机理图

6.4　本章小结

本章采用了溶剂热法制备 MIL-101@TiO$_2$ 复合材料,通过调控反应物中 TiF$_4$ 的含量、溶剂热反应温度和反应时间,获得完整的 MIL-101@TiO$_2$ 八面体颗粒。将优化后制备的 MIL-101@TiO$_2$ 碳化形成 Cr$_2$O$_3$/C@TiO$_2$ 核壳结构复合材料。MIL-101(Cr)不仅是 TiO$_2$ 壳体生长的基底,也是 Cr$_2$O$_3$ 和 Cr-TiO$_2$ 中 Cr 元素的来源。MIL-101(Cr)中少量的 Cr^{3+} 掺杂到 TiO$_2$ 中导致 TiO$_2$ 带隙值降低从而拓宽了其光响应区域。复合材料中由金属有机骨架中获得的碳加速了电子和空穴的分离。Cr$_2$O$_3$/C@TiO$_2$ 独特的纳米结构和元素组合赋予了复合材料优异的光催化性能。在 0.4 V(vs.Ag/AgCl)的工作电压和 300 W 氙灯的照射下,光电流密度达到 2.1 mA/cm^2,约为纯 TiO$_2$ 的 3.5 倍。全幅光照下,Cr$_2$O$_3$/C@TiO$_2$ 的最大析氢速率为 446 μmol/(h·g),约为纯二氧化钛的 4 倍。可见光照射下(λ>420 nm),最大析氢速率可以达到 25.5 μmol/(h·g)。Cr$_2$O$_3$/C@TiO$_2$ 的光催化性能优于 TiO$_2$,这主要是由于其较宽的可见光响应范围和光生电子-空穴对的有效分离,本书为设计和制备高效光催化体系提供了有用的信息。

参考文献

[1] LIN L H,WANG C,REN W,et al.Photocatalytic overall water splitting by conjugated semiconductors with crystalline poly(triazine imide) frameworks [J].Chemical science,2017,8(8):5506-5511.

[2] FU C F,WU X J,YANG J L.Material design for photocatalytic water splitting from a theoretical perspective[J].Advanced materials,2018,30 (48):1802106.

[3] XU F Y,ZHU B C,CHENG B,et al.1D/2D TiO$_2$/MoS$_2$ hybrid nanostructures for enhanced photocatalytic CO$_2$ reduction [J]. Advanced optical materials,2018,6(23):1800911.

[4] ZHU Z,KAO C T,TANG B H,et al.Efficient hydrogen production by photocatalytic water-splitting using Pt-doped TiO$_2$ hollow spheres under visible light[J].Ceramics international,2016,42(6):6749-6754.

[5] DE BRITO J F,TAVELLA F,GENOVESE C,et al.Role of CuO in the

modification of the photocatalytic water splitting behavior of TiO_2 nanotube thin films[J].Applied catalysis B:environmental,2018,224:136-145.

[6] PAN L,MUHAMMAD T,MA L,et al.MOF-derived C-doped ZnO prepared via a two-step calcination for efficient photocatalysis[J].Applied catalysis B:environmental,2016,189:181-191.

[7] HSU C C,WU N L.Synthesis and photocatalytic activity of ZnO/ZnO_2 composite[J].Journal of photochemistry and photobiology A:chemistry,2005,172(3):269-274.

[8] LI J,WU D D,IOCOZZIA J,et al.Achieving efficient incorporation of π-electrons into graphitic carbon nitride for markedly improved hydrogen generation[J].Angewandte chemie,2019,131(7):2007-2011.

[9] YUAN Y J,SHEN Z K,WU S T,et al.Liquid exfoliation of $g\text{-}C_3N_4$ nanosheets to construct 2D-2D $MoS_2/g\text{-}C_3N_4$ photocatalyst for enhanced photocatalytic H_2 production activity[J].Applied catalysis B:environmental,2019,246:120-128.

[10] FU J W,YU J G,JIANG C J,et al.$G\text{-}C_3N_4$-based heterostructured photocatalysts[J].Advanced energy materials,2018,8(3):1701503.

[11] WANG X C,MAEDA K,THOMAS A,et al.A metal-free polymeric photocatalyst for hydrogen production from water under visible light[J].Nature materials,2009,8(1):76-80.

[12] LIANG Q,CUI S N,LIU C H,et al.Construction of CdS@UiO-66-NH_2 core-shell nanorods for enhanced photocatalytic activity with excellent photostability[J].Journal of colloid and interface science,2018,524:379-387.

[13] HUANG H M,DAI B Y,WANG W,et al.Oriented built-in electric field introduced by surface gradient diffusion doping for enhanced photocatalytic H_2 evolution in CdS nanorods[J].Nano letters,2017,17(6):3803-3808.

[14] LI G Q,SHEN Q Y,YANG Z Z,et al.Photocatalytic behaviors of epitaxial $BiVO_4$(010) thin films[J].Applied catalysis B:environmental,2019,248:115-119.

[15] ZHAO X,HU J,YAO X,et al.Clarifying the roles of oxygen vacancy in W-doped $BiVO_4$ for solar water splitting[J].ACS applied energy mate-

rials,2018,1(7):3410-3419.

[16] LI X Y,PI Y H,XIA Q B,et al.TiO$_2$ encapsulated in Salicylaldehyde-NH$_2$-MIL-101(Cr) for enhanced visible light-driven photodegradation of MB[J].Applied catalysis B:environmental,2016,191:192-201.

[17] LI L H,YU L L,LIN Z Y,et al.Reduced TiO$_2$-graphene oxide heterostructure as broad spectrum-driven efficient water-splitting photocatalysts[J].ACS applied materials and interfaces,2016,8(13):8536-8545.

[18] XU M,CHEN Y,MAO G B,et al.Tio$_2$ nanoparticle modified α-Fe$_2$O$_3$ nanospindles for improved photoelectrochemical water oxidation[J]. Materials express,2019,9(2):133-140.

[19] CHEN Y Z,LI A X,LI Q,et al.Facile fabrication of three-dimensional interconnected nanoporous N-TiO$_2$ for efficient photoelectrochemical water splitting[J].Journal of materials science and technology,2018,34(6):955-960.

[20] WANG H,HU X T,MA Y J,et al.Nitrate-group-grafting-induced assembly of rutile TiO$_2$ nanobundles for enhanced photocatalytic hydrogen evolution[J].Chinese journal of catalysis,2020,41(1):95-102.

[21] WANG P,XU S Q,CHEN F,et al.Ni nanoparticles as electron-transfer mediators and NiS$_x$ as interfacial active sites for coordinative enhancement of H$_2$-evolution performance of TiO$_2$[J].Chinese journal of catalysis,2019,40(3):343-351.

[22] SHEN J,WANG R,LIU Q Q,et al.Accelerating photocatalytic hydrogen evolution and pollutant degradation by coupling organic cocatalysts with TiO$_2$[J].Chinese journal of catalysis,2019,40(3):380-389.

[23] ZHANG Y J,MAO F X,WANG L J,et al.Recent advances in photocatalysis over metal-organic frameworks-based materials[J].Solar RRL,2020,4(5):1900438.

[24] MENG A Y,ZHANG L Y,CHENG B,et al.Dual cocatalysts in TiO$_2$ photocatalysis[J].Advanced materials,2019,31(30):1807660.

[25] LI H J,ZHOU Y,TU W G,et al.State-of-the-art progress in diverse heterostructured photocatalysts toward promoting photocatalytic per-

formance[J].Advanced functional materials,2015,25(7):998-1013.

[26] PAN C,JIA J,HU X Y,et al.In situ construction of g-C_3N_4/TiO_2 heterojunction films with enhanced photocatalytic activity over magnetic-driven rotating frame[J].Applied surface science,2018,430:283-292.

[27] HE F,MENG A Y,CHENG B,et al.Enhanced photocatalytic H2-production activity of WO_3/TiO_2 step-scheme heterojunction by graphene modification[J].Chinese journal of catalysis,2020,41(1):9-20.

[28] ZHANG W,ZHANG H W,XU J Z,et al.3D flower-like heterostructured TiO_2@Ni(OH)$_2$ microspheres for solar photocatalytic hydrogen production [J].Chinese journal of catalysis,2019,40(3):320-325.

[29] WANG B X,WANG Z Q,CUI Y J,et al.Cr_2O_3@TiO_2 yolk/shell octahedrons derived from a metal-organic framework for high-performance lithium-ion batteries[J].Microporous and mesoporous materials,2015,203:86-90.

[30] DAI B Y,YU Y R,CHEN Y K,et al.Construction of self-healing internal electric field for sustainably enhanced photocatalysis[J].Advanced functional materials,2019,29(16):1807934.

[31] ZHANG F,JIN T,ZENG R C,et al.Cr_2O_3 nanoparticles modified TiO_2 nanotubes for enhancing visible photoelectrochemical performance[J]. Journal of nanoscience and nanotechnology,2014,14(9):7022-7026.

[32] MAO G B,XU M,YAO S Y,et al.Direct growth of Cr-doped TiO_2 nanosheet arrays on stainless steel substrates with visible-light photoelectrochemical properties[J].New journal of chemistry,2018,42(2): 1309-1315.

[33] IRIE H,SHIBANUMA T,KAMIYA K,et al.Characterization of Cr(Ⅲ)-grafted TiO_2 for photocatalytic reaction under visible light[J].Applied catalysis B:environmental,2010,96(1/2):142-147.

[34] JUN T H,LEE K S.Cr-doped TiO_2 thin films deposited by RF-sputtering[J].Materials letters,2010,64(21):2287-2289.

[35] WANG J,LV Y H,ZHANG Z H,et al.Sonocatalytic degradation of azo fuchsine in the presence of the Co-doped and Cr-doped mixed crystal TiO_2 powders and comparison of their sonocatalytic activities [J]. Journal of hazardous materials,2009,170(1):398-404.

[36] LU Y, ZHANG H C, CHAN J Y, et al. Homochiral MOF-polymer mixed matrix membranes for efficient separation of chiral molecules [J]. Angewandte chemie, 2019, 131(47): 17084-17091.

[37] FENG Y, CHEN Q, JIANG M Q, et al. Tailoring the properties of UiO-66 through defect engineering: a review [J]. Industrial and engineering chemistry research, 2019, 58(38): 17646-17659.

[38] DANG S, ZHU Q L, XU Q. Nanomaterials derived from metal-organic frameworks [J]. Nature reviews materials, 2018, 3: 17075.

[39] GUO S H, ZHANG P C, FENG Y, et al. Rational design of interlaced Co_9S_8/carbon composites from ZIF-67/cellulose nanofibers for enhanced lithium storage [J]. Journal of alloys and compounds, 2020, 818: 152911.

[40] XU H L, YIN X W, ZHU M, et al. Constructing hollow graphene nanospheres confined in porous amorphous carbon particles for achieving full X band microwave absorption [J]. Carbon, 2019, 142: 346-353.

[41] YANG W P, LI X X, LI Y, et al. Applications of metal-organic-framework-derived carbon materials [J]. Advanced materials, 2019, 31(6): 1804740.

[42] CAI Z X, WANG Z L, KIM J, et al. Hollow functional materials derived from metal-organic frameworks: synthetic strategies, conversion mechanisms, and electrochemical applications [J]. Advanced materials, 2019, 31(11): 1804903.

[43] WANG C C, WANG X, LIU W. The synthesis strategies and photocatalytic performances of TiO_2/MOFs composites: a state-of-the-art review [J]. Chemical engineering journal, 2020, 391: 123601.

[44] FENG Y, LU H Q, GU X L, et al. ZIF-8 derived porous N-doped ZnO with enhanced visible light-driven photocatalytic activity [J]. Journal of physics and chemistry of solids, 2017, 102: 110-114.

[45] CHANG N, ZHANG H, SHI M S, et al. Regulation of the adsorption affinity of metal-organic framework MIL-101 via a TiO_2 coating strategy for high capacity adsorption and efficient photocatalysis [J]. Microporous and mesoporous materials, 2018, 266: 47-55.

[46] ALAMELU K, RAJA V, SHIAMALA L, et al. Biphasic TiO_2 nanoparticles decorated graphene nanosheets for visible light driven photocatalytic degradation of organic dyes [J]. Applied surface science, 2018, 430:

145-154.

[47] STANKOVICH S, DIKIN D A, PINER R D, et al. Synthesis of graphene-based nanosheets via chemical reduction of exfoliated graphite oxide[J]. Carbon, 2007, 45(7): 1558-1565.

[48] YUAN R R, YUE C L, QIU J L, et al. Highly efficient sunlight-driven reduction of Cr(Ⅵ) by TiO_2@NH_2-MIL-88B(Fe) heterostructures under neutral conditions[J]. Applied catalysis B: environmental, 2019, 251: 229-239.

[49] MONIZ S J A, SHEVLIN S A, MARTIN D J, et al. Visible-light driven heterojunction photocatalysts for water splitting: a critical review[J]. Energy and environmental science, 2015, 8(3): 731-759.

[50] PENG Y H, HUANG G F, HUANG W Q. Visible-light absorption and photocatalytic activity of Cr-doped TiO_2 nanocrystal films[J]. Advanced powder technology, 2012, 23(1): 8-12.

[51] JOY J, MATHEW J, GEORGE S C. Nanomaterials for photoelectrochemical water splitting-review[J]. International journal of hydrogen energy, 2018, 43(10): 4804-4817.

[52] TU W G, XU Y, YIN S M, et al. Rational design of catalytic centers in crystalline frameworks[J]. Advanced materials, 2018, 30(33): 1707582.

[53] ZHANG Y F, QIU L G, YUAN Y P, et al. Magnetic Fe_3O_4@C/Cu and Fe_3O_4@CuO core-shell composites constructed from MOF-based materials and their photocatalytic properties under visible light[J]. Applied catalysis B: environmental, 2014, 144: 863-869.

[54] WEI X J, ZHANG Y H, HE H C, et al. Carbon-incorporated NiO/Co_3O_4 concave surface microcubes derived from a MOF precursor for overall water splitting[J]. Chemical communications (Cambridge, England), 2019, 55(46): 6515-6518.

[55] JIAO Z B, CHEN T, XIONG J Y, et al. Visible-light-driven photoelectrochemical and photocatalytic performances of Cr-doped $SrTiO_3$/TiO_2 heterostructured nanotube arrays[J]. Scientific reports, 2013, 3: 2720.